T0393511

History of Analytic Philosophy

More information about this series at
http://www.palgrave.com/gp/series/14867

Anna Drabarek · Jan Woleński ·
Mateusz M. Radzki
Editors

Interdisciplinary Investigations into the Lvov-Warsaw School

palgrave
macmillan

Editors
Anna Drabarek
Institute of Philosophy and Sociology
The Maria Grzegorzewska University
Warsaw, Poland

Mateusz M. Radzki
Institute of Philosophy and Sociology
The Maria Grzegorzewska University
Warsaw, Poland

Jan Woleński
Department of Social Sciences
University of Information Technology
and Management
Rzeszów, Poland

History of Analytic Philosophy
ISBN 978-3-030-24485-9 ISBN 978-3-030-24486-6 (eBook)
https://doi.org/10.1007/978-3-030-24486-6

Cover illustration: Grzegorz Gajewski/Alamy Stock Photo

This Palgrave Macmillan imprint is published by the registered company Springer Nature Switzerland AG
The registered company address is: Gewerbestrasse 11, 6330 Cham, Switzerland

Series Editor's Foreword

During the first half of the twentieth century, analytic philosophy gradually established itself as the dominant tradition in the English-speaking world, and over the last few decades it has taken firm root in many other parts of the world. There has been increasing debate over just what 'analytic philosophy' means, as the movement has ramified into the complex tradition that we know today, but the influence of the concerns, ideas and methods of early analytic philosophy on contemporary thought is indisputable. All this has led to greater self-consciousness among analytic philosophers about the nature and origins of their tradition, and scholarly interest in its historical development and philosophical foundations has blossomed in recent years, with the result that history of analytic philosophy is now recognized as a major field of philosophy in its own right.

The main aim of the series in which the present book appears, the first series of its kind, is to create a venue for work on the history of analytic philosophy, consolidating the area as a major field of philosophy and promoting further research and debate. The 'history of analytic philosophy' is understood broadly, as covering the period from the last three decades of the nineteenth century to the start of the twenty-first century, beginning with the work of Frege, Russell, Moore and Wittgenstein, who are generally regarded as its main founders, and the influences upon them, and going right up to the most recent developments. In allowing the 'history' to extend to the present, the aim is to encourage engagement with contemporary debates in philosophy, for example, in showing how the concerns of early analytic philosophy relate to current concerns. In focusing on analytic philosophy, the aim is not to exclude comparisons with other—earlier or contemporary—traditions, or consideration of figures or themes that some might regard as marginal to the analytic tradition but which also throw light on analytic philosophy. Indeed, a further aim of the series is to deepen our understanding of the broader context in which analytic philosophy developed, by looking, for example, at the roots of analytic philosophy in neo-Kantianism or British idealism, or the connections between analytic philosophy and phenomenology, or discussing the work of philosophers who were important in the development of analytic philosophy but who are now often forgotten.

One group of philosophers who indeed played a significant role in the development of analytic philosophy but whose work is relatively unknown in the English-speaking world is the Lvov-Warsaw School, founded by Kazimierz Twardowski (1866–1938). Its beginning can be dated to 1895, when Twardowski became Professor of Philosophy at the University of Lvov, and it lasted until 1939, when the outbreak of the Second World War effectively brought its activities to an end. Its members included philosophers and logicians whose work is now recognized in contemporary philosophy, such as Jan Łukasiewicz (1878–1956), Stanisław Leśniewski (1886–1939) and Alfred Tarski (1901–1983), but there were many more who contributed to the formation of a distinctive analytic subtradition. While focusing on questions of logic and methodology, their interests were much more wide-ranging than is commonly appreciated, as the present volume, edited by Anna Drabarek, Jan Woleński and Mateusz M. Radzki, demonstrates. Entitled *Interdisciplinary Investigations into the Lvov-Warsaw School*, it is interdisciplinary not only in approaching the School from an interdisciplinary perspective but also in bringing out the interdisciplinarity of the School itself.

This volume complements an earlier book published in this series in 2013, *Studies in the History and Philosophy of Polish Logic*, which was edited by Kevin Mulligan, Katarzyna Kijania-Placek and Tomasz Placek. This earlier book had indeed focused on logic and philosophy of logic. So I am delighted that we are now publishing a sequel that places the Lvov-Warsaw School, which forms the core of the Polish analytic tradition, in broader historical context. Jan Woleński, to whom the earlier book was dedicated, in recognition of all the work he has done over the years in promoting Polish analytic philosophy, outlines this historical context in his contribution to the present volume. His chapter, in the first part on 'History, culture and axiology', also contains chapters on the fate of Polish philosophers during the Second World War, the views that members of the Lvov-Warsaw School had about the role of universities and about ethical values, and the various interrelationships between the people and texts involved. The second part comprises chapters on psychology, and the third part chapters on logic and methodology, which includes a chapter on metaphilosophy and the connections to the ideas of 'multidisciplinarity', 'interdisciplinarity' and 'transdisciplinarity'. Not only in this chapter but also in the volume as a whole, as Anna Drabarek says at the end of her introductory chapter, the volume thus makes a contribution to contemporary debate about interdisciplinarity in philosophy and science. As far as its contribution to this series on the history of analytic philosophy is concerned, it also shows the importance and value of understanding the broader historical context of the various schools and subtraditions that make up the very complex movement that we now know as analytic philosophy. There should no longer be any excuse for ignoring the richness and significance of Polish analytic philosophy, and I hope the present volume will inspire further work on the issues and questions that the members of the Lvov-Warsaw School addressed.

Berlin, Germany
May 2019

Michael Beaney

Preface

Just like history and culture can never be finally closed and finished, neither can philosophical reflection. The non-final nature of the results of philosophical work is not a defect or a shortfall, but the very essence of reflections which are aimed at pursuing truth. This is even more distinctly visible in the continuous and repeated process of the new being inspired by the old. The consonance of that which was achieved in philosophy with that which is currently analysed by various philosophers is the subject matter of analyses contained in this volume, devoted to interdisciplinary investigation into the Lvov-Warsaw School.

Philosophy pursued by Kazimierz Twardowski and his students founded its activities, still continued today, on openness and interdisciplinarity. They were fully aware that disregard for transformations taking place in science entails negative consequences for the philosophical understanding of reality. It is to Twardowski and his students that we owe the establishment of a centre of analytical philosophical thought in Lvov at the beginning of the twentieth century. Their approach to philosophical problems, clarity of thought and language, a specific type of rationalism and realism provided the foundations for this school of philosophy in Poland and warranted its international success, particularly in the logic of methodology.

This volume, entitled *Interdisciplinary Investigations into the Lvov-Warsaw School*, presents a broad spectrum of topics which may be grouped into three content clusters: (1) History, Culture and Axiology; (2) Psychology; (3) Logic and Methodology.

Warsaw, Poland — Anna Drabarek
Rzeszów, Poland — Jan Woleński
Warsaw, Poland — Mateusz M. Radzki

Contents

1	**Introduction** *Anna Drabarek*	1

Part I History, Culture and Axiology

2	**Lvov-Warsaw School: Historical and Sociological Comments** *Jan Woleński*	17
	1 Introduction	17
	2 The Beginnings	18
	3 Two Periods of the School's History	22
	4 The Method	25
	5 The Fate of LWS Members	28
	References	33

3 **The Victims and the Survivors: The Lvov-Warsaw School and the Holocaust** 35
Elżbieta Pakszys
1 Introduction 35
2 The Victims 37
3 The Survivors 39
4 Conclusion 43
References 44

4 **The Lvov-Warsaw School on the University and Its Tasks** 47
Włodzimierz Tyburski
1 Introduction 47
2 The Ethos of a University Professor 48
3 Kazimierz Twardowski and the Modern University 51
4 Tadeusz Czeżowski on the Tasks of the University 53
5 Conclusion 59
References 61

5 **The Axiology Project of the Lvov-Warsaw School** 63
Anna Drabarek
1 Introduction 63
2 Objectivism in the Understanding of Good 65
3 The Category of Act 72
4 The Category of Happiness 77
5 Conclusion 82
References 83

6 **Interpersonal and Intertextual Relations in the Lvov-Warsaw School** 87
Anna Brożek
1 Introduction 87
2 Influence and Its Necessary Conditions 88
2.1 Contact Between Philosophers 88
2.2 Types of Influence 89
2.3 Resonance: Its Spheres, Nature and Manifestations 91
2.4 Identification of Influence 92

3 Influences Within the LWS: Examples 93
3.1 Contacts Between Twardowski and His Students 94
3.2 Twardowski's Resonance in the Didactic Sphere 96
3.3 Examples of Twardowski's Theoretical Resonance in His Disciples 97
4 Theoretical Resonance of Twardowski's Disciples in Himself 106
5 Diversity of the LWS and Its Sources 108
References 111

Part II Psychology

7 The Relationship Between Judgments and Perceptions from the Point of View of Twardowski's School 119
Stepan Ivanyk
1 Introduction 119
2 Twardowski 120
3 Students of Twardowski 128
4 Conclusion 137
References 138

8 On the Lvov School and Methods of Psychological Cognition 141
Teresa Rzepa
1 Introduction 141
2 Background 144
3 Methods of Psychological Cognition 149
4 In Lieu of a Conclusion 155
References 156

9 The Interdisciplinary Nature of Władysław Witwicki's Psychological Investigations 159
Amadeusz Citlak
1 Introduction 159
2 The Theory of Cratism and Alfred Adler's Individual Psychology 160

3 'Wiara oświeconych', 'Dobra Nowina wg Mateusza i Marka' and the Dorpat School of the Psychology of Religion 163
3.1 The Problem of Suppositions and Cognitive Dissonance 164
4 Cratic Portraits and Psychobiographies of Socrates and Jesus Christ 170
References 173

Part III Logic and Methodology

10 Pragmatic Rationalism and Pragmatic Nominalism in the Lvov-Warsaw School 179
Witold Marciszewski
1 Introduction 179
2 On How to Do Science: Controversies About Nominalism in the Lvov-Warsaw School 181
2.1 What Does It Mean to Be Indispensable? Ockham's Maxim in a Modern Interpretation 181
2.2 The Puzzle of Existence: What Exists in the Domain of Physics? 184
2.3 A Pragmatically Liberalised Variant of Nominalism 186
3 Abstraction as the Driving Force of Scientific Progress 189
3.1 Some Comments on the Relationship Between Rationalism and Nominalism 189
3.2 Ontological Types as Ones to Be Addressed from a Syntactic Point of View 190
3.3 Syntactic Types and Their Relationship to Ontological Types According to CG 191
3.4 On How the Operator of Abstraction Transforms an Incomplete Expression into a Complete One to Name an Abstract Entity 194
3.5 On Abstract Constituents of Physical Reality 196

3.6 The Issue of Correspondence Between Syntactic and Ontological Types: Its Philosophical Import 197
4 Conclusion: On the Progress of Science Owing to Ever Higher Levels of Abstraction 198
References 202

11 Some Problems Concerning Axiom Systems for Finitely Many-Valued Propositional Logics 205
Mateusz M. Radzki
1 Introduction 205
2 Many-Valued Logics of Łukasiewicz and the Functionally Complete Three-Valued Logic 206
3 The Rosser-Turquette Method 208
4 Conclusion 211
References 214

12 The Methodological Status of Paraphrase in Selected Arguments of Tadeusz Kotarbiński and Kazimierz Ajdukiewicz 217
Marcin Będkowski
1 Introduction 217
2 Reistic Paraphrase 218
3 Explicative Paraphrase 225
4 Conclusion 234
References 236

13 The Metaphilosophical Views of Zygmunt Zawirski Against the Background of Contemporary Discussions on Interdisciplinarity in Science 239
Jarosław Maciej Janowski
1 Introduction 239
2 The Metaphilosophical Views of Zawirski 242
3 Multidimensionality of the Conceptual Framework Related to the Postulate of Interdisciplinarity 251
4 Conclusion 255
References 256

14 The Informational Worldview and Conceptual Apparatus 259
Paweł Stacewicz
1 Introduction 259
2 Between Worldview and Philosophy 260
3 The Informational Conceptual Apparatus 266
4 Informational Worldview 269
5 A Discussion of Marciszewski's Optimistic Realism 272
References 278

Index 281

Notes on Contributors

Marcin Będkowski Assistant Professor, University of Warsaw, Faculty of Philosophy and Sociology, Warsaw, Poland.

Anna Brożek Professor of Philosophy, University of Warsaw, Faculty of Philosophy and Sociology, Warsaw, Poland.

Amadeusz Citlak Assistant Professor, Institute of Psychology, Polish Academy of Sciences, Warsaw, Poland.

Anna Drabarek Professor of Philosophy, The Maria Grzegorzewska University, Institute of Philosophy and Sociology, Warsaw, Poland.

Stepan Ivanyk Assistant Professor, Kazimierz Twardowski Philosophical Society of Lviv, Lviv, Ukraine.

Jarosław Maciej Janowski Assistant Professor, The Maria Grzegorzewska University, Institute of Philosophy and Sociology, Warsaw, Poland.

Witold Marciszewski Professor of Philosophy, Foundation for Computer Science, Logic and Formalized Mathematics in Warsaw, Poland.

Elżbieta Pakszys Professor of Philosophy, Adam Mickiewicz University, Poznań, Poland.

Mateusz M. Radzki Assistant Professor, The Maria Grzegorzewska University, Institute of Philosophy and Sociology, Warsaw, Poland.

Teresa Rzepa Professor of Psychology, University of Social Science and Humanities, Department of Psychology in Poznań, Warsaw, Poland.

Paweł Stacewicz Assistant Professor, Warsaw University of Technology, Faculty of Administration and Social Sciences, Warsaw, Poland.

Włodzimierz Tyburski Professor of Philosophy, The Maria Grzegorzewska University, Institute of Philosophy and Sociology, Warsaw, Poland.

Jan Woleński Professor of Philosophy, University of Information Technology and Management, Department of Social Sciences, Rzeszów, Poland.

List of Figures

Chapter 12

Fig. 1 The argument for the methodological thesis of reism, where 'C' stands for the 'conclusion', and 'P' stands for the 'premise' 219

Fig. 2 The argument for the thesis of idealism in the semantic formulation proposed by Ajdukiewicz, where 'C' stands for the 'conclusion', and 'P' stands for the 'premise' 227

List of Tables

Chapter 11

Table 1 The definitions of the propositional connectives $\sim, \rightarrow, \vee, \wedge, \leftrightarrow$ in the three-valued logic of Łukasiewicz (In the truth-tables for binary propositional connectives, the truth-value of the first argument is given in the vertical line, the truth-value of the second argument is given in the horizontal line, and the outcome is given in the intersection of these lines) 207

Table 2 Słupecki's *T*-function 208

Chapter 12

Table 1 A one-to-one correspondence between constituents of the thesis of idealism and its paraphrase 232

1
Introduction

Anna Drabarek

The aim of this book is to present the heritage of the Lvov-Warsaw School from both the historical and the philosophical perspective. The historical view is focused on the beginnings and the dramatic end of the School brought about by the outbreak of World War II. The philosophical view, on the other hand, encompasses a broad spectrum of issues, including logical, epistemological, axiological, and psychological problems.

It should be emphasised that the philosophical perspective reveals the interdisciplinary nature of studies carried out by Kazimierz Twardowski and his students. The School's achievements range across philosophy, ethics, psychology, logic, and mathematics. Their unifying factor consists in the fact that they all reflect the rationalist and analytical trait of the School's philosophy, characteristic of the Polish philosophy at the beginning of the twentieth century.

A. Drabarek (✉)
Institute of Philosophy and Sociology,
The Maria Grzegorzewska University, Warsaw, Poland
e-mail: adrabarek@aps.edu.pl

A. Drabarek et al. (eds.), *Interdisciplinary Investigations into the Lvov-Warsaw School*, History of Analytic Philosophy,
https://doi.org/10.1007/978-3-030-24486-6_1

This volume, entitled 'Interdisciplinary Investigations into the Lvov-Warsaw School', consists of three parts: (1) History, Culture, and Axiology; (2) Psychology; (3) Logic and Methodology.

The first part begins with a paper by Jan Woleński, entitled 'The Lvov-Warsaw School: Historical and Sociological Comments'. The author introduces Twardowski as the founder of an analytical philosophical school in Lvov at the end of the nineteenth century. Twardowski, as a charismatic teacher, did not take long to train a group of young philosophers. This group became the Lvov-Warsaw School just after the end of World War I. Although some philosophers of the Lvov-Warsaw School were active until the end of the twentieth century (and even at the beginning of the twenty-first century), the School itself existed as an organised and coherent scientific project only until 1939. Its end was brought about by World War II and the ensuing political changes.

Woleński points out that the Lvov-Warsaw School was a large and complex community (of about 80 persons) spreading across several generations and circles. Twardowski and his earliest students (including Jan Łukasiewicz, Stanisław Leśniewski, Władysław Witwicki, Tadeusz Kotarbiński, Kazimierz Ajdukiewicz, Zygmunt Zawirski, and Tadeusz Czeżowski) constituted the first generation of the Lvov-Warsaw School. They trained the second generation, acting mostly in Lvov and Warsaw. The Warsaw School of Logic with Leśniewski, Łukasiewicz, and Alfred Tarski as its main representatives, became the most famous branch of the Lvov-Warsaw School, working on various problems of mathematical logic. This circle had close connections with the Polish School of Mathematics. Philosophers, like Ajdukiewicz, Czeżowski, Kotarbiński or Zawirski, contributed to epistemology, ontology, the philosophy of language and philosophy of science. On the other hand, the Lvov-Warsaw School was also active in the history of philosophy, ethics, aesthetics, and psychology. Thus, all basic parts of philosophy were represented in the Lvov-Warsaw School. Due to the multi-ethnic character of the Polish society in the interwar period, the Lvov-Warsaw School consisted not only of Poles (the majority), but also Jews and Ukrainians.

Woleński considers the significance of the Lvov-Warsaw School from two perspectives. Its logicians achieved many fundamental results which influenced logical studies around the world. Some of them (Tarski's theory of truth, Łukasiewicz's many-valued logic, Leśniewski's systems) were important for general philosophy (in particular, Tarski's works in semantics resulting in a change in the philosophical orientation of logical empiricism from syntax-oriented to semantic-oriented). Taking into account the role of the Lvov-Warsaw School in Poland, it made an essential contribution to the country's philosophical culture.

The next essay, written by Elżbieta Pakszys, 'The Victims and the Survivors: the Lvov-Warsaw School and the Holocaust', discusses the subject of death in general and in particular as it concerns philosophers of the Lvov-Warsaw School during the Holocaust perpetrated by the Nazi Germans during World War II.

Pakszys reports on individual stories of survivors rescued from prisons and concentration camps, as well as stories of those who managed to survive the Holocaust in hiding, revealed in the course of post-war studies. All of them survived concentration camps and resumed active lives after the war. They all shared the experience of being imprisoned in the female concentration camp at Ravensbrück.

'The Lvov-Warsaw School on the University and Its Tasks' by Włodzimierz Tyburski presents the ideas and assertions of key representatives of the Lvov-Warsaw School concerning the role and tasks of the university and the ethical duties (deontology) of the scholar. The discussion of these topics is preceded by an exposition of selected ideas on the role and mission of the university shared by outstanding Polish scholars from the fifteenth through the twentieth century. This issue was of utmost importance to Twardowski, founder of the Lvov-Warsaw School, as well as his followers, including Czeżowski, Kotarbiński, Ajdukiewicz, Witwicki and others.

In his essay, 'The Lvov-Warsaw School on the University and Its Tasks', Tyburski presents the ideas and analyses concerning the above topics carried out by Twardowski and Czeżowski, since they are highly representative of the entire Lvov-Warsaw School. All of its representatives shared their conviction that the university is committed to (a) serving the ideal of truth, (b) harmoniously combining its research, educational, and teaching functions, (c) educating creative individuals displaying intellectual, aesthetic as well as moral qualities, (d) contributing to cultural development, (e) influencing both its immediate and more distant surroundings by applying the results of its research activity. Czeżowski, Twardowski and all other representatives of the Lvov-Warsaw School firmly stood for the university's independence, freedom of scientific research, and independence from governments and politics.

Another hallmark of the School was a common catalogue of duties involved in the education and training of university students. Tyburski emphasises that the university as proposed by the Lvov-Warsaw School differs considerably from the contemporary proposals, where the university is designed as an enterprise offering teaching and research services defined by the current demand. These are two very different institutions, with different structures, functions, goals, and following entirely different standards.

'The Axiology Project of the Lvov-Warsaw School' by Anna Drabarek is concerned with the discussion of values by Twardowski, Witwicki, Kotarbiński, Czeżowski, Władysław Tatarkiewicz, and Karol Frenkel. Drabarek claims that they developed an interesting and inspiring axiology project, being a creative development of the postulate of ethics as a science, and emphasising the importance of facts in the substantiation of norms. Drabarek points out that the axiology project of the Lvov-Warsaw School, contrary to David Hume's paradigm, removes the chasm between facts and values.

Firstly, Drabarek claims that an attempt at showing the need for a constant confrontation of facts and values becomes the most important challenge for morality. This project also had a pragmatic objective, as it was one of the ways of restoring the meaning of life to the Polish nation which, after 120 years of slavery, was only beginning to rebuild its identity. It could therefore be called a kind of therapy through meaning, emphasising the ability to shape man's existence based on the values of truth and good.

Secondly, Drabarek considers the problem of objectivism in the understanding of good found in the philosophy of Tatarkiewicz and Czeżowski. Good is considered an absolute value, one that is unqualified and objective. It may be defined as a property of things, or seen as a way in which things exist. Good can be known intuitively, but intuition is understood as an autonomous kind of knowledge, or a certain experience which may be improved.

Drabarek then goes on to analyse the category of act found in the philosophy of Twardowski, Witwicki, Frenkel, and Kotarbiński. Every rule of a just act must contain an empirical element and be based on an inductive juxtaposition of individual instances of just acts or individual rules. Despite the absolute nature of good, rules are always relative. When judging an act, one should take into account the person's character, or the permanent direction of their will, and through introspection we can understand and judge the dynamics of human action.

Finally, Drabarek presents the category of happiness found in the philosophy of Twardowski, Tatarkiewicz, Kotarbiński, and Czeżowski. Drabarek discusses the personal role model of a reliable guardian, or an honest man who comes to the aid of all those who need help. Moreover, Drabarek points out the heroism which results from a specific treatment of values in the process of moral improvement.

In the paper 'Interpersonal and Intertextual Relations in the Lvov-Warsaw School', Anna Brożek starts with the premise that two kinds of relations are essential to any philosophical school. Firstly, there are interpersonal relations between its members, including those between teachers and students. Secondly, there are intertextual relations between elements of the school's output, namely between problems, conceptual schemes, methods, or theories accepted by its various representatives.

The chapter consists of two parts. In the first part, the conceptual scheme for the analysis of various types of interpersonal and intertextual relations is provided. The author claims that contacts between philosophers (oral or written, unilateral or bilateral) are a necessary, but not a sufficient element of one philosopher's influence on another. Influence takes place if contacts between philosophers result in some actions either of them takes, or convictions they embrace. Such influence may be positive or negative. Moreover, one may be more or less aware of being influenced by someone else.

The second part of the chapter provides some examples of interpersonal and intertextual relations within the Lvov-Warsaw School, particularly as regards contacts between Twardowski, the School's founder, and his direct students. Further on, some examples of Twardowski's influence on the first generation of the Lvov-Warsaw School are provided, and the way his influence was revealed is described. In addition, important characterological and theoretical differences between Twardowski's students are discussed.

In the conclusion, a hypothesis is proposed that the characteristic nature of interpersonal and intertextual relations within the Lvov-Warsaw School were caused by Twardowski's methodology and his interdisciplinary approach to philosophical problems.

The second part of this volume, devoted to psychology, begins with a text by Stepan Ivanyk on 'The Relationship Between Judgments and Perceptions from the Point of View of Twardowski's School'. Ivanyk emphasises that 'judgment' and 'perception' were undoubtedly two of the most important terms used in psychological and logical theories developed in Twardowski's School. The issue of the mutual relationship between the correlates of these terms has been a concern for many of the School's representatives for several decades. No comprehensive study of analyses they have performed in this respect has appeared in the philosophical literature, however. The main aim of this chapter is to fill this gap by pursuing the following specific objectives.

Ivanyk reviews the evolution of views held by the School's representatives on the relationship between judgments and perceptions. He divides the School's representatives who investigated the relationship between judgments and perceptions into two groups: (1) those who tried to determine the place of judgments and perceptions in the overall structure of the psyche by seeking to answer the question 'Are judgments a kind of perception (or vice versa)?' and 'Are judgments an element of perception (or vice versa?)'. Their views evolved from the relationship of '*genus proximum-definiendum*' to that of co-existence or a 'part-to-whole' relationship between judgment and perception; (2) those who tried to determine the justifying power of perceptions for the truthfulness of judgments related to them. The result of research carried out by these representatives of the School was rejection of the possibility of a logical justification of judgments by relevant perceptions, and conclusion about a cause-and-effect relationship between them.

In the next chapter, 'On the Lvov School and Methods of Psychological Cognition', Teresa Rzepa emphasises that the Polish academic psychology derives from the descriptive (Franz Brentano) as well as the experimental and physiological psychology (Wilhelm Wundt), which prevailed in Europe at the end of the nineteenth century. These trends were encountered by the first Lvov psychologists—Witwicki, Stefan Baley, Stefan Błachowski, Mieczysław Kreutz—who chose their academic paths consistently following the founder of the Lvov-Warsaw School, Kazimierz Twardowski.

The chapter focuses on a presentation of their preferred methods of psychological cognition. Rzepa shows that introspection was considered to be the direct and obvious source of knowledge about mental life, as sensations and experiences are always 'someone's' and are felt by 'someone'. Attempts were made to mitigate the constraints of introspection by applying the experimental method and the object method, consisting in the interpretation of facts of somebody else's mental life. These methods, relying on Twardowski's theory of actions and products, constituted the basis on which the Lvov psychological laboratory was established.

One of the principles followed in its work was the conservative and critical approach to the test-mania that infested the world in the first decades of the twentieth century. This saved Poland from being flooded with tests, and at the same time changed the attitude of academic psychologists to psychological practice, and initiated actions aimed at establishing the psychologist profession and defining the model of communication between a psychologist and his or her patients. Methodological achievements of the Lvov School of Psychology also include a skilful application of the analytical method according to the idea that any phenomenon should first be clearly, explicitly, and reliably described before it can be explained.

Rzepa concludes that the result of the methodological soundness of the Lvov School was the ethos consisting of an inquisitive approach to the subject and tools of psychological cognition.

In the last chapter, 'The Interdisciplinary Nature of Władysław Witwicki's Psychological Investigations', Amadeusz Citlak presents the work of Witwicki—one of Twardowski's closest students. Citlak points out that Witwicki's psychological achievements are still little known in international psychological literature. It is quite surprising given that his research was very original and could very well be included in the mainstream of the most important theoretical problems of psychology in the first half of the twentieth century. That this did not happen was mainly due to political reasons (restrictions imposed on the School and its discontinuation after World War I) and ideological trends (after 1918, Polish psychologists published mainly in the Polish language).

In this chapter, Citlak draws attention to the most original achievements of Witwicki and their similarity to the then contemporary achievements of world psychology. They involved a number of areas, in both the theoretical and the empirical tradition. Firstly, Alfred Adler's individual psychology and his theory of will to power. Citlakpoints out that several years earlier Witwicki had developed his original theory of kratism, which is similar to the assumptions of the Austrian psychologist, and yet remains virtually unknown in world literature. Secondly, the Dorpat School of the Psychology of Religion, whose achievements coincided in some respects with Witwicki's research into religious beliefs. Thirdly, the theory of cognitive dissonance developed by Leon Festinger, and fourthly—based on the theory of kratism– the psychobio—aphies of Socrates (already in 1909) and Jesus Christ (1958), which are two of the earliest non-psychoanalytical psychobiographies in the world.

Citlak concludes that Witwicki's psychological research was interdisciplinary and concerned with important problems of contemporary psychology. It was also part of the historical (or cultural-historical) psychology postulated by Wundt. One of the most important features of his analyses was the use of description and interpretation of empirical data, and to a lesser extent the experiment or methods derived from the natural sciences.

In the first chapter of the third part of this volume, 'Pragmatic Rationalism and Pragmatic Nominalism in the Lvov-Warsaw School', Witlod Marciszewski considers the stance of 'pragmatic realism' concerning the existence of abstract objects. He points out that the acknowledgement of abstracts may be motivated either by their direct intellectual vision, as claimed by Plato, or by their 'indispensability' for scientific progress, as claimed by Willard V. O. Quine and a number of other famous thinkers. For such a practical approach, the term 'pragmatic rationalism' should be in order ('pragmatic Platonism' is used more frequently, but in the present context this term would be misleading).

Its followers included some eminent members of the Lvov-Warsaw School: Ajdukiewicz with his empirical methodology of sciences, and

Tarski in view of the abundant use of highly abstract concepts in his mathematical research.

Radical nominalism pioneered by Leśniewski and Kotarbiński, denying the existence of classes, did not comply with the methods and results of sciences. Marciszewski points out that there is a moderate version of nominalism which pragmatically admits the existence of classes for the sake of scientific progress. This version has been suggested by Andrzej Grzegorczyk under the name of 'liberal reism'.

In this pragmatic form, nominalism can comply with pragmatic rationalism, and the other way around. The latter is not bound to deny what nominalism says about the ontological and epistemological priority of physical individuals. Even a rationalist as ardent as Kurt Gödel shared with nominalists (unlike Plato) the view that observational statements concerning physical individuals are most basic in the structure of our knowledge.

An example of such reconciliation is found in Tarski's paper entitled 'Foundations of the Geometry of Solids'. Tarski abandons Euclid's idea of the point as a primitive concept, in accordance with the claim of reistic nominalism that only solids can play such a role. Instead, he starts from the concept of a sphere, defining 'the point as the class of all spheres which are concentric with a given sphere'.

Marciszewski concludes that the typically abstract concept of class (attacked by radical nominalism) proves indispensable to do justice to the ontological priority of physical individuals—a hallmark of nominalism. This way, while doing justice to nominalism, at the same time Tarski scores considerable points for pragmatic rationalism as well. Such constructive collaboration of the two opposing camps within the School sheds new light on the School's ability to solve difficult problems through a vigorous and 'penetrative discussion' between opposite standpoints.

In the next chapter in this part, 'Some Problems Concerning Axiom Systems for Finitely Many-Valued Propositional Logics', Mateusz M. Radzki examines some problems related to the axiomatisation of finitely many-valued propositional logics. As he points out, it has recently been demonstrated that neither a particular axiom system for the functionally complete three-valued logic nor a certain general method of constructing axiom systems for finitely many-valued logics satisfies some salient metalogical requirements.

Firstly, Radzki examines an axiom system for the functionally complete three-valued logic based on the well-known Mordchaj Wajsberg axiom system for the three-valued logic of Łukasiewicz. The examined axiom system was introduced by Jerzy Słupecki in 1936. Radzki demonstrates that this axiom system is not semantically complete.

Then, Radzki investigates a method of constructing axiom systems for standard many-valued propositional logics introduced by John B. Rosser and Atwell R. Turquette. The Rosser-Turquette method is considered to be a solution to the problem of the axiomatisability of a particular class of many-valued logics, i.e., standard many-valued propositional logics, including the finitely many-valued propositional logics of Łukasiewicz and Emil Post.

However, Radzki demonstrates that the Rosser-Turquette method fails to produce adequate axiom systems for a particular class of many-valued propositional logics, i.e., for the finitely many-valued propositional logics of Łukasiewicz. Investigations presented in his essay concern the mutual definability of the Rosser-Turquette standard connectives and connectives of the finitely many-valued propositional logics of Łukasiewicz. The conclusion is that every Rosser-Turquette axiom system for finitely many-valued propositional logics of Łukasiewicz that satisfies the necessary condition of being a sound system is semantically incomplete.

In 'The Methodological Status of Paraphrase in Selected Arguments of Tadeusz Kotarbiński and Kazimierz Ajdukiewicz', Marcin Będkowski investigates so-called arguments from paraphrase introduced both by Kotarbiński and Ajdukiewicz. These arguments assert as one of the premises that there is a paraphrastic relationship between some expressions. This type of argumentation is quite common in analytic philosophy, perhaps even essential for it. Meanwhile, Będkowski writes, the methodological status of both the operation and the resulting relationship requires further reflection.

The chapter discusses two arguments from paraphrase: Kotarbiński's argument for the methodological thesis of reism, and Ajdukiewicz's argument against transcendental idealism. Będkowski shows the place of paraphrastic premises in the structure of selected philosophical arguments, reconstructs the implied meaning of 'the relationship of paraphrase', and presents reasons supporting those premises.

Będkowski discusses the argument for the methodological thesis of reism which is based upon a weak formulation of the semantic thesis and the methodological principle of ontological parsimony. A reist should be able to formulate a reistic theory of paraphrase, and in particular should not assume non-reistic expressions to be primarily meaningful. It seems that his theory should match non-reistic and reistic sentences on the sole basis of their syntactic structure. As stated by Kotarbiński himself, the justification supporting the semantic thesis of reism is naively intuitive and based on common induction.

As Będkowski points out, Ajdukiewicz tries to reformulate the position of idealism, expressed in the thesis of idealism. To this end, he establishes a correlation between expressions belonging to the idealist language and expressions belonging to the language of semantic epistemology. By using metalogical theorems, he demonstrates the falsehood of the thesis of idealism in its semantic formulation. However, Będkowski concludes, the principles applied in the procedure of semantic paraphrase used by Ajdukiewicz are not—and probably cannot be—strictly defined, which undermines the validity of his account.

In 'The Metaphilosophical Views of Zygmunt Zawirski Against the Background of Contemporary Discussions on Interdisciplinarity in Science', Jarosław Maciej Janowski examines the meta-philosophical views of Zawirski. The author presents them against the background of the meta-philosophic views of the entire Lvov-Warsaw School.

Janowski discusses the application of Zawirski's meta-philosophical concepts as exemplified in his analyses regarding the concept of time presented in one of his most important works, entitled 'L'évolution de la notion du temps'. In the chapter, Zawirski's meta-philosophic views are compared with the concepts of 'multidisciplinarity', 'interdisciplinarity', and 'transdisciplinarity' used in contemporary meta-theoretical discussions concerning the philosophy of science.

The aim of the chapter is to examine Zawirski's philosophical reflections and meta-philosophical views for correspondence to contemporary postulates regarding the conduct of interdisciplinary research in science and philosophy. In relation to the three concepts mentioned above, Janowski emphasises that Zawirski's views were in line with the idea of multidisciplinarity and interdisciplinarity. Janowski believes, however, that Zawirski would have been cautious about the idea of interdisciplinarity, particularly when it comes to the postulate of integrating all disciplines of science into one super-discipline. And he would have been just as sceptical about the idea of transdisciplinarity in science.

The last chapter, 'The Informational Worldview and Conceptual Apparatus' by Paweł Stacewicz, relates to the thought of the Lvov-Warsaw School in a most contemporary way, as it refers to computer science. Despite this contemporary context, it addresses rather general issues, representative of the School's history. These are: (1) the relationship between philosophy and worldview, (2) the relationship between philosophy and exact sciences, (3) the realistic cognitive optimism concept developed by Marciszewski.

The author begins with declarations made by selected representatives of the School (including Twardowski and Tatarkiewicz), and then specifies the relationship between worldview and philosophy as feedback occurring over time. He explains that, on the one hand, socially grounded worldviews contribute to the development of philosophy (including the creation of new doctrines), and on the other hand, sufficiently well-developed philosophical systems shape individual and social worldviews. At the same time, following some works by Ajdukiewicz, Stacewicz makes a distinction between worldview proper (which is a collection of beliefs about the world and life in it) and a certain grid of notions that provide for a precise categorisation of the world ('worldview perspective').

Next, Stacewicz describes the conceptual apparatus of computer science (consisting of such terms as information, computing, algorithm, computability, and uncomputability), and then explains some preliminary assumptions of the informational worldview. These are: (1) Each being has a certain information content; (2) The human mind is an information processing system; (3) With the development of human civilization, the complexity of problems solved by the mind (through information processing) keeps increasing.

The author then moves on to discuss the philosophical concept developed by Marciszewski (one of the School's contemporary continuators), which refers to assumption (3). He calls this concept 'cognitive optimism' for short, or, in a more descriptive way, 'realistic optimism about understanding and transforming the world'. It says that creativity of the human mind is inexhaustible, which means that the mind is able to 'follow infinitely' the growing complexity of problems in the world. In the mathematical-computational convention, Marciszewski's key thesis reads as follows: 'For each theory and for each model of computation there is a possibility of such enrichment or transformation of the model/theory that problems which have not been solved yet become resolvable (on a new basis)'.

Stacewicz counters this thesis with three 'computational' arguments: (1) The proposed strengthening or transformation of a model/theory requires reference to the actual infinity; (2) We have no proof that models stronger than the Turing model of computation are physically feasible; (3) Some computations/algorithms used to solve uncomputable problems (in the case of the Turing model) are not fully controllable by man (e.g. because of randomness). Due to the existence of such arguments, the question of cognitive optimism about the potential of computational techniques is left open.

This volume does not only discuss the history of philosophy represented by the Lvov-Warsaw School. It also includes a reflection on the condition of contemporary philosophy from the perspective of concepts developed by its representatives.

Moreover, the studies presented in this book are related to problems of contemporary science and interdisciplinarity as its distinctive characteristic. This volume is, therefore, not only a collection of interdisciplinary analyses of the Lvov-Warsaw School philosophy, but also an investigation into the interdisciplinarity of science and philosophy itself.

Part I

History, Culture and Axiology

2

Lvov-Warsaw School: Historical and Sociological Comments

Jan Woleński

1 Introduction

Władysław Tatarkiewicz wrote:

> After defending my doctoral thesis, I went to Lvov; I wanted to see how Poles worked. I discovered immediately that their working style was different, and better; Twardowski's school taught how to work in a scientific way... Unfortunately, that was in the summer of 1910; due to Ukrainian riots lectures were suspended, and I only had enough time to attend two lectures and two classes taught by the master. I do not remember what they were about any more, but I do remember the method: I liked it better than those of other Western professors. (Tatarkiewicz and Tatarkiewicz 1979: 125)

And here is a passage from Roman Ingarden:

J. Woleński (✉)
Department of Social Sciences, University of Information Technology and Management, Rzeszów, Poland
e-mail: jan.wolenski@uj.edu.pl

A. Drabarek et al. (eds.), *Interdisciplinary Investigations into the Lvov-Warsaw School*, History of Analytic Philosophy,
https://doi.org/10.1007/978-3-030-24486-6_2

> I wrote this paper in the winter of 1919/1920, prompted to do so by the situation in Polish philosophy I saw in Warsaw after I arrived there in the summer of 1919. This is not the proper place to describe the rather sorry state of affairs I witnessed then. Suffice it to say that I was struck first of all by the almost complete unawareness of what was going on in contemporary Western European philosophy. I believed something should be done about it, to the extent of my possibilities. I had just recently come back from the large academic centre which Göttingen was at the beginning of the century, having spent several years listening to Husserl and the phenomenologists. (Ingarden 1919: 269)

Two outstanding philosophers, both direct witnesses of what was going on in Polish philosophy in the years 1910–1920, and yet they leave us with such dissimilar impressions. Both were up to date with developments in the international philosophical community. Ingarden wrote about it explicitly, yet Tatarkiewicz had not come from a philosophical backwater either, but from Marburg, one of the two main centres of Neo-Kantianism. The reasons for this difference were twofold: scientific (they belonged to two different philosophical camps, even though Tatarkiewicz was probably the first author in Poland to write about phenomenology), and personal (Ingarden suggested on multiple occasions that his academic career had been thwarted by Kazimierz Twardowski and some of his students), but the question about who was right is a legitimate one. For me it is a rhetorical question, as I have written more than once that the Lvov-Warsaw School (hereinafter referred to as LWS) was one of the most important, if not the most important event in the history of Polish philosophical thought. This chapter is yet another expression of this belief.

2 The Beginnings

The beginnings of LWS go back to 1895, when Twardowski became professor at the University of Lvov. He was a student of Franz Brentano and took over his metaphilosophical programme. Its main points were as follows: (1) The method of philosophy is the same as that of other empirical sciences, and therefore the philosopher is required to put forward clear and well-substantiated theses; (2) The basis of philosophy is descriptive psychology which investigates not so much the genesis of psychic phenomena but rather their content; (3) The characteristic feature of psychic phenomena is their intentionality, i.e., the fact they are directed at an object; (4) Philosophy should avoid speculative issues. Twardowski added the following elements: (a) he linked psychological analysis with semiotics, for example by analysing the properties of representations by considering the nature of their corresponding names; (b) he introduced an important distinction between the content and object of representation, which had been systematically mixed up before; (c) he distinguished acts from products, which was to help limit psychologism; (d) he strongly emphasised the role of clarity and precision in philosophical thinking; (e) he performed a critical analysis (very influential in Poland) of the relativist understanding of truthfulness; (f) while he recommended minimalism and was sceptical about building large philosophical systems, he did not shy away from traditional issues of metaphysics, epistemology and axiology (particularly ethics); (g) he believed that philosophy should not be influenced by worldview.

The beginnings of Twardowski's work in Lvov were not easy. The Lvov University, though founded in 1652 by King John II Casimir, had not fully developed until 1818. It was a provincial school, mostly German-speaking. Philosophy was taught there on a very elementary (still scholastic) level. The liberalisation that came with the Habsburg monarchy changed the character of the Lvov University. Austrian authorities wanted it to be dualist (utraquist)—Polish-Ukrainian. This project only partially succeeded, as Ukrainians did not have enough faculty members, and the school was quickly polonised. This should be borne in mind, as the University's status was a bone of contention in the Polish-Ukrainian conflict (mentioned by Tatarkiewicz in the quote above) even in the interwar period.

Polonisation of the Lvov University did not have any immediate academic consequences. Before Twardowski, virtually no philosophy was taught there. The beginnings of his work were difficult. Here is one reminiscence (Witwicki 1920: XI):

> He found lecture halls to be nearly empty. A few acquaintances…, some bolder foreign students came to look in, some out of politeness, some out of curiosity, to see what the young professor looked like and how he taught. Gradually, the hall began to fill up and soon there was not enough room for all who wanted to come, so the lectures had to be moved outside of University walls, because none of its assembly halls could accommodate all of the students who rushed early in the morning to find a sitting place. (Witwicki 1920: XI)

Twardowski made it quite clear that he intended to establish a philosophical school in Poland. The poor condition of philosophy in Lvov before he came to the city helped further his plans. It would certainly have been much more difficult to accomplish what he intended in a more developed and established philosophical community, such as Cracow, for example. As he was beginning from scratch, he could mould philosophy in Lvov in line with his designs.

After 10 years, Władysław Witwicki recalls, crowds of students attended his lectures and he had to organise entrance exams for those who wanted to attend his seminar. Lvov was gaining renown in the Polish philosophical community. Tatarkiewicz must have known, somehow, that it was worthwhile going to Lvov to see how Poles worked. In any case, quite soon after Twardowski was appointed professor, the name Twardowski's School or the Lvov School came into use. Even though it was based in Lvov, it could be called a Polish school since Twardowski's students enrolled from all parts of the partitioned country.

Twardowski had a very clear-cut view of the place of Polish philosophy in the international context. Henryk Struve, considered to be a doyen of Polish philosophy, wrote an editorial for the first issue of 'Ruch Filozoficzny' ('Philosophical Movement', Struve 1911). He held the philosophical capacities of Poles in high esteem. He believed they were destined to create a universal synthesis drawing on the tradition of Polish national philosophy, as long as they ensured its continuity. Twardowski's opinion on the matter (Twardowski 1911) was different. In particular, he examined the relationship of Polish philosophy and the thought developed in the leading countries, i.e. Germany, France and Great Britain. He believed there were two options, either for the Polish philosophy to fence of (with a 'Chinese wall', he used to say) from what was going on in the world, or to toe the line with one of the dominant philosophies. He believed neither of these options was good. The former would eventually result in breaking any contacts with recent developments in global philosophy; the latter—in Polish philosophy melting in with some foreign thought. According to Twardowski, Polish philosophers should look for new and valuable ideas in many different places (which would help avoid both the Chinese wall effect and the consequences of submitting to any foreign philosophy), and try to process them in an original way.

He repeated these ideas (Twardowski 1918), postulating, for example, that textbooks for Polish students, particularly in the history of philosophy, should not be translated from foreign languages but written by Polish authors. He pointed out that a German, English or French historian of philosophy would always focus on the thought of his own nation, understating the achievements of others, and furthermore—leave out entirely the philosophies of smaller nations. The well-known textbook 'Historia Filozofii' ('A History of Philosophy') by Tatarkiewicz (first published in 1931) is a practical application of this programme. Other textbooks inspired by Twardowski's project include Witwicki's 'Psychologia' ('Psychology') (1925–1927) and 'Elementy teorii poznania, logiki formalnej i metodologii nauk' ('Elements of the Theory of Cognition, Formal Logic and Methodology of Sciences') by Tadeusz Kotarbiński (1929).

Twardowski was of the opinion that effective philosophical life must develop within a clear-cut organisational framework. He contributed to the founding of 'Przegląd Filozoficzny' ('Philosophical Review'—a magazine created by Władysław Weryha at the end of the nineteenth century; the first issue was published in 1898), he founded the Polish Philosophical Society in Lvov (in 1904) and the first laboratory of experimental psychology (in 1907); in 1911, he started the publication of 'Philosophical Movement', already mentioned above, a magazine providing information on philosophical life (books, magazines, organisations, congresses, conferences etc.) at home and abroad (to help Polish philosophers follow developments in the philosophical life worldwide); he initiated Polish philosophical conventions (three were held in the years 1918–1939—Lvov in 1923, Warsaw in 1927, and Cracow in 1936); and participated in the works of a number of educational institutions—he was an ardent advocate of the education of women. He placed a lot of emphasis on the propaedeutics of philosophy in secondary schools.

3 Two Periods of the School's History

The history of LWS can be divided into two periods: the years 1895–1918 (the Lvov period) and the years 1918–1939 (the Lvov-Warsaw period); the latter could be extended until 1945, though for obvious reasons the years 1939–1945 were very special. It is worth noting that the name 'Lvov-Warsaw School' (or, more precisely 'l'école de Lvov et Varsovie') was probably first used by Kazimierz Ajdukiewicz in his opening address to participants in the Congress of Scientific Philosophy held in Paris in 1935. It did not come into popular use until after World War II.

Whether or not LWS existed after 1945 is disputable. I have consulted this matter with several persons. Izydora Dąmbska categorically claimed that LWS existed until 1939, or until 1945 at the most. Still, she might have been influenced by personal resentments. Marian Przełęcki and Klemens Szaniawski, students of Twardowski's disciples, also claimed that LWS had not survived World War II. In their opinion, World War II changed the nature of philosophy, as it could not be separated from worldview as radically as had been postulated by Twardowski. Jerzy Pelc, another philosophical grandchild of the founder of LWS, once told me: 'It is a pity you did not include us [i.e. himself, Przełęcki and Szaniawski] in the School'. Some authors refer to a new, or third, generation of the LWS (Jadacki and Paśniczek 2006) as a continuation of the tradition started by Twardowski. In my opinion, such concept is not well-founded. An analogy with the Vienna Circle is instructive in this respect. The Circle did not survive Austria's 'Anschluss' into the Third Reich and did not reconstruct after World War II. Its representatives, including such prominent figures as Rudolf Carnap, continued their work and views. Nevertheless, if the Vienna

Circle still existed, it was only in the individual achievements of its representatives. The same applies to LWS. Even though the political factor (the regime prevalent in Poland after 1945) certainly did not contribute to renewing LWS as a coherent and well-organised philosophical group, it is not certain whether its reconstruction would have taken place had the pre-war political realities subsisted.

Twardowski's main task in the first period of the School's history was to train professional philosophers. His first students were Jan Łukasiewicz and Witwicki. The former specialised in logic, while the latter, though dealing mainly in psychology (he developed the so-called theory of kratism), made a great contribution to Polish philosophy by translating Plato and authoring important works in ethics, aesthetics and the philosophy of religion.

Both of them very soon began lecturing at Lvov University. In the years 1907–1914, Twardowski and his collaborators mentioned above educated (though one should remember that Twardowski was the main figure in this process) a group of philosophers. The best known of them were (in alphabetical order): Kazimierz Ajdukiewicz, Tadeusz Czeżowski, Tadeusz Kotarbiński and Zygmunt Zawirski, but we should also add Hersch Bad, Bronisław Bandrowski, Marian Borowski, Daniela Gromska (née Tenner), Salomon Igel, Stanisław Kaczorowski, Kazimierz Sośnicki, Franciszek Smolka, and Michał Treter, among others. In addition, two outstanding philosophers came to Lvov who had not studied under Twardowski's direction, namely Stanisław Leśniewski and Władysław Tatarkiewicz, who has already been mentioned here; the former came to write his doctoral thesis, the latter—his post-doctoral dissertation.

The way Twardowski understood philosophy emphasised issues in semiotics, logic and the methodology of sciences. Many of Twardowski's disciples specialised in these areas. Apart from Łukasiewicz, they included Ajdukiewicz, Czeżowski, Kaczorowski, Kotarbiński, Leśniewski and Zawirski, as well as others. Another strong group was that of psychologists (along with Witwicki it included Stefan Baley, Stefan Błachowski, Mieczysław Jaxa-Bykowski, Mieczysław Kreutz)—psychology was already beginning to detach from philosophy. Nevertheless, all philosophical sciences were represented in the period discussed here: ethics, aesthetics and the history of philosophy. Twardowski's seminars were also attended by linguists (e.g. Jerzy Kuryłowicz), classical philologists (Ryszard Ganszyniec), Romanists (Zygmunt Czerny), Germanists (Zygmunt Łempicki), historians of literature (Juliusz Kleiner, Manfred Kridl), historians of culture (Stanisław Łempicki). The group gathering around Twardowski was large and diversified. It would not be an exaggeration to say the humanities in Lvov were largely dominated by Twardowski's students.

The logical orientation, which began to dominate already in the Lvov period, grew in strength even more in the years 1918–1939 when the School transformed from Lvov to Lvov-Warsaw. It was then that the Warsaw school of logic was formed. It was developed by Leśniewski and Łukasiewicz, who became professors at the University of Warsaw (reactivated in 1915) at the Department of Mathematics and Physics. The Warsaw School of Logic was a shared achievement of mathematicians and philosophers. It was related to the programme of developing Polish mathematics (proposed by Zygmunt Janiszewski), based on the premise that the set theory, topology and the basics of mathematics, i.e. areas closely related to mathematical logic, should be the main areas of study. The first disciple of Leśniewski and Łukasiewicz was Alfred Tarski, who was to become the most prominent Polish logician and probably also the best-known representative of the LWS. The Warsaw School of Logic also included: Stanisław Jaśkowski, Adolf Lindenbaum, Andrzej Mostowski, Mojżesz Presburger, Jerzy Słupecki, Bolesław Sobociński and Mordchaj Wajsberg (shortly before the war they were joined by Henryk Hiż and Czesław Lejewski). Kotarbiński became professor of philosophy at the University of Warsaw in 1919 and began lecturing there. His students (and those of Tatarkiewicz) included, among others: Jan Drewnowski, Janina Hosiasson-Lindenbaumowa, Mieczysław Geblewicz, Mieczysław Milbrandt, Jan Mosdorf, Maria Niedźwiedzka-Ossowska, Stanisław Ossowski, Antoni Pański, Edward Poznański, Jakub Rajgrodzki, Dina Sztejnbarg (later Janina Kotarbińska), Mieczysław Wallis-Walfisz and Aleksander Wundheiler. Eventually, also Tatarkiewicz moved to Warsaw (he also lectured in Poznań and Vilnius), just like Witwicki. Twardowski remained in Lvov; the other Lvov professor was Ajdukiewicz (from 1928; in the years 1926–1928 he lectured in Warsaw). Their students included: Walter Auerbach, Leopold Blaustein, Izydora Dąmbska, Eugenia Ginsberg-Blausteinowa, Maria Kokoszyńska-Lutmanowa, Seweryna Łuszczewska-Romahnowa, Henryk Melhberg, Zygmunt Schmierer, Witwicki and Stefan Swieżawski. Apart from the Warsaw School of Logic, the great majority of the above-mentioned philosophers, which could be referred to as the second generation of the LWS, specialised in semantics and the methodology of sciences, even though it also included e.g. ethicists (Ossowska), aestheticians (Ossowski), and historians of philosophy (Swieżawski).

Although the LWS was based in two cities, Lvov and Warsaw, its representatives appeared at other Polish universities as well, for example Czeżowski in Vilnius (those influenced by him included, among others: Edward Csató, Antoni Korcik, Jan Rutski, as well as theoreticians of law—Sawa Frydman (after World War II—Czesław Nowiński) and Józef Zajkowski); Zawirski in Poznań (from 1928; from 1937—in Cracow; one of his Poznań students was Zbigniew Jordan). Those related to the Warsaw group included Józef M. Bocheński (who lived in Switzerland), and Jan Salamucha (who later became professor of theology in Cracow). Together with Drewnowski and Sobociński, in 1936 they founded the Cracow Circle whose main goal was to apply mathematical logic to neo-scholastic problems. Of all Polish universities, only the Catholic University of Lublin did not have an LWS representative in the interwar period.

4 The Method

In 1939, the School founded by Twardowski included at least eighty academics involved in creative work. This group included not only philosophers in the strict sense of the word, but also representatives of other specialities—philologists, psychologists, linguists, etc. It may be referred to as LWS in the broadest sense. Despite there being all reason to allow for blurred divisions between humanities, such a broadly-conceived group could hardly be considered a philosophical school. The group included more than sixty philosophers, which is a lot. In fact, they formed one of the largest schools of philosophy in the world—if not the largest one. The main criterion determining the identity of any school is, naturally, the views and interests of its members. In addition, the person of the founder is taken into account, usually the first teacher in a particular circle, and the location. The second and third criterion (genetic and geographical, respectively) is relatively simple in the case of LWS (see below). Nearly all of Twardowski's students and their students considered themselves to belong to the School. LWS was also clearly based in its two main centres. There were some exceptions, however. Bocheński was not a student of Lvov-Warsaw professors, and yet he considered himself to be a member of the School. Ingarden started as a student of Twardowski, later became professor in Lvov, and yet he cannot be considered a member of the LWS. Leon Chwistek was a logistician, thus representing one of the main LWS specialties, a professor in Lvov, but not a representative of Twardowski's School. In fact, he was one of its radical critics.[1] Another example is Benedykt Bornstein, who wrote his doctoral thesis under Twardowski's guidance, but did not feel connected to his supervisor's tradition.

If we consider views and interests, the situation is not simple. This was explicitly pointed out by Dąmbska:

> The Lvov philosophers [the same could be said about Warsaw philosophers – J.W.] did not share any common doctrine, any consistent worldview. Their spiritual community was based not on the content, but on the method, the way of doing philosophy, and a shared scientific language. Which is why the School was able to produce: spiritualists and materialists, nominalists and realists, logisticians and psychologists, philosophers of nature and theoreticians of art. (Dąmbska 1948: 17)

We might say, applying today's metaphilosophical benchmarks, that the method consisted in applying philosophical analysis, which

[1]Alan R. Perreiah (1991) once faulted me for not including Chwistek in the LWS. He failed to notice, however, that being a logistician in Lvov was not enough to belong to the School. Having remarked on Chwistek's critical approach to the LWS, let me also note that other critics (and there were more of them; see Woleński 1997) included Henryk Elzenberg (see Zegzuła-Nowak 2017), Ludwik Fleck and Florian Znaniecki.

is why LWS became part of analytical philosophy alongside such currents as neo-positivism or the philosophy of colloquial language. The domains of philosophy one wanted to specialise in, views held on particular issues, the attitude to religion or political beliefs did not matter much. Which is why LWS included representatives of all philosophical sciences, advocates of very disparate views: both materialists and idealists, atheists and people of faith (Bocheński repeatedly emphasised that he was more an analytical philosopher than a Thomist), as well as people of both right and left wing orientation. Also the philosophical influences impacting LWS were diverse. Next to Brentano, already mentioned above, and some of his students, for example Alexius Meinong, they included Gottlob Frege, Edmund Husserl (contrary to Ingarden's opinion cited above), French conventionalists, particularly Henri Poincaré, Bertrand Russell, George E. Moore, and then the Vienna Circle.

One of the research areas explored by the LWS, namely logic, was particularly emphasised, which contributed to the School often being perceived as a logical school.[2] It should be remembered, however, that logic in the narrower sense (formal, symbolic, mathematical) is distinguished from logic in the broader sense, which also includes semiotics and the methodology of sciences. The Warsaw School of Logic was concerned with logic in the narrower sense (which greatly contributed to its being one of the most important schools worldwide), and the same applies, to a certain extent, to Ajdukiewicz, Czeżowski, Zawirski or Zygmunt Schmierer. Łukasiewicz believed that mathematical logic was neither part of philosophy nor mathematics, but an autonomous domain of science. Today, this view is no longer upheld. There is no reason for excluding the works of Warsaw logisticians from philosophy, much less than those of the other philosophers mentioned above. This suggests LWS should be more specifically described as including those who creatively practiced logic, first of all the Warsaw School of

[2]I cannot help but recount the following story here. I had read the history of analytic philosophy written by an American author (I will leave out the name and the title). I only found one Polish name in it—Alfred Tarski. I asked the author why he had left out Polish philosophers. He answered that he knew there were logisticians in Poland, but had never heard of any philosophers.

Logic and those who were rather philosophers than logisticians, but whose style of doing philosophy was largely influenced by logic—such as Ajdukiewicz, Czeżowski, Kotarbiński and Zawirski from the first generation of Twardowski's students, as well as Bocheński, Dąmbska, Drewnowski, Hosiasson-Lindenbaumowa, Jordan, Kokoszyńska, Kotarbińska, Mehlberg, Poznański, Salamucha, Schmiererand Wundheiler. It dealt mostly with semiotics and the methodology of sciences, but many of them also considered classical problems of ontology or epistemology.

The borderline between the logical wing of the LWS and the rest is fluid. Twardowski himself was not concerned with formal logic, he pointed out the dangers of overly trusting formalisation, which he referred to as symbol-mania. Ossowski, who eventually became a sociologist, and Ossowska, who worked mainly in the field of descriptive ethics, published important works in semiotics, Tatarkiewicz wrote papers on the general methodology of sciences, and Swieżawski (a neo-Thomist who always emphasised his connection to Twardowski and Ajdukiewicz)—on the history of philosophy. The link between all of them was undoubtedly the analytical method, and in terms of general philosophical views—the stance which Ajdukiewicz referred to as anti-irrationalism, meaning that only those statements should be treated seriously which were intersubjectively verifiable and intersubjectively communicable, and only those should be accepted which have been verified. Indeed, it was Brentano's line continued by Twardowski.

5 The Fate of LWS Members

I should nevertheless share some of my doubts concerning the personal composition of the LWS. When writing my monograph (Woleński 1985, 1989), I consulted the personal issues with Dąmbska. She insisted that the list of LWS members should be as broad as possible and follow the genetic criterion. She stressed that one of the criteria of belonging, not only on the part of professional philosophers, was a strong subjective sense of being part of Twardowski's school. Based on my talks with Dąmbska, I compiled a list (Woleński 1985: 338–339) which includes such names as Edward Csató, Irena Filozofówna, Milbrandt, Mosdorf and Szumowski. The information I have gathered about them does not suggest that they had any particular sense of connection to the LWS tradition. Nevertheless, applying all criteria, at least 70 names can certainly be included in the LWS in the broadest sense. Another problem concerns the nationality of LWS members. The fact that about 15% of them were of Jewish descent was naturally very well known. A research carried out by Stepan Ivanyk (Ivanyk 2014) shows that Twardowski had quite a few students of Ukrainian origin, not only Baley (which had been known beforehand), but also Stepan Oleksyuk, Gabriel Kostelnyk and Miron Zarycki. This fact is interesting in view of the many Polish–Ukrainian conflicts, including academic ones, e.g. concerning the creation of a Ukrainian university in the Borderland.

Two outstanding representatives of the LWS died shortly before the war: Twardowski in 1938, and Leśniewski in 1939. The war dealt a tremendous blow at the LWS with the deaths of Auerbach, Bad, Blaustein, Blausteinowa, Hosiasson-Lindenbaumowa, Igel, Lindenbaum, Zygmunt Łempicki, Milbrandt, Mosdorf, Ortwin, Pański, Presburger, Rajgrodzki, Rutski, Salamucha, Schmierer, Treter, Wajsberg, and Zajkowski (many of whom were of Jewish descent), i.e. at least 25% of the overall number of LWS representatives (in the broadest sense). In addition, an unknown number of students and alumni, who were just then about to begin their scholarly work, died as well. Difficult conditions and a persistent sense of threat did not prevent LWS members from actively participating in underground education at the secondary and tertiary level (Ajdukiewicz and Dąmbska in Lvov, Kotarbiński, Łukasiewicz, Hiż, Mr and Mrs Ossowski, Salamucha, Sobociński and Tatarkiewicz in Warsaw, Czeżowski in Vilnius, Zawirski in Cracow, and others).[3] Underground students included, among others, Jan Kalicki (a logistician, in a sense the last of the LWS alumni), Andrzej Grzegorczyk, Helena Rasiowa and Klemens Szaniawski. Zawirski died in 1948. Many of the LWS

[3]Some also actively participated in the resistance movement, e.g. Hiż worked as a cryptologist at the Home Army (AK) Headquarters, and Sobociński was part of the political leadership of the National Armed Forces (NSZ).

philosophers had left the country shortly before or during the war, or soon after 1945. They included: Bocheński (who stayed mainly abroad), Hiż, Kalicki, Czesław Lejewski, Jordan, Łukasiewicz, Poznański, Sobociński, Tarski and Wundheiler.

Nearly a half of pre-war LWS philosophers remained in Poland and joined the effort to rebuild and organise academic life in the country. After World War II, two universities—in Lvov (the cradle of the LWS) and Vilnius were no longer within Polish borders. New universities were founded—in Lublin (a state-run one, as the Catholic University of Lublin (KUL) had already existed), Łódź, Toruń and Wrocław. This resulted in migrations in the academic community, also among philosophers. Kotarbiński and Tatarkiewicz remained in Warsaw (Kotarbiński was also a professor in Łódź), Zawirski in Cracow; but others moved to new locations—Ajdukiewicz to Poznań, then Warsaw, Czeżowski to Toruń, Kokoszyńska to Wrocław, Korcik to Lublin (KUL), and Dąmbska to Cracow (from 1957). Also Polish emigrants worked effectively abroad, including Tarski at Berkeley (where he created a large California School of Logic), Łukasiewicz in Dublin, Lejewski in Manchester, Mehlberg in Chicago and Toronto, Hiż in Philadephia, Jordan in Ottawa, or Sobociński at Notre Dame. Contribution to academic life both in Poland and abroad is the first historical achievement of the LWS, and applies not only to the years 1900–1939, but the subsequent period as well.

Until 1948, academic life in Poland followed mostly its pre-war course, but was then faced with the ideological offensive of Marxism and administrative restrictions. The so-called bourgeois philosophers were no longer allowed to teach at all (e.g. Tatarkiewicz and Dąmbska), others had severely limited teaching hours (e.g. Mr and Mrs Ossowski), still others could only teach logic (e.g. Ajdukiewicz, Czeżowski, Kotarbiński). In the years 1951–1952, Marxists strongly attacked prominent LWS philosophers (as well as Ingarden or Catholic philosophers), in particular Twardowski, Ajdukiewicz and Kotarbiński. Despite this criticism and pressure, not one of LWS representatives (or representatives of other so-called bourgeois orientations) changed their views (except for some isolated instances). Those professors who had been removed resumed their lectures after 1956, and polemics on the Marxist side became academic. Most of the philosophers educated in Poland after 1945 were directly or indirectly influenced by the LWS, not only by its views, but also, in some cases mostly, by its way of understanding philosophy and its tasks. It was the LWS (as well as other interwar philosophers) who contributed to the development of a philosophical culture that was distinctly different than in other Eastern bloc countries. From the second half of the 1970 s, it was more important (in statistical terms, of course, not without exceptions) that one was a philosopher, than whether one was a Marxist or a non-Marxist. This is yet another historical achievement of the LWS.

The history of philosophy in Poland has a long and uninterrupted tradition dating back to 1400, i.e. when the Cracow Academy was re-established as a fully-fledged University. In fact, its history represents the longest uninterrupted tradition in the Balkans to Scandinavia belt. Nevertheless, Polish philosophical thought always lagged behind western philosophy, to a greater or lesser extent. In fifteenth century Cracow, all scholastic currents were still present when *via moderna* already prevailed all over Europe. The Renaissance in Poland began later than in Western countries, though earlier than in Scandinavia, for example. And even though the Polish school of international law (a slightly exaggerated name, but never mind) was well-known and even influential, and the same applies to conciliarism or the Arian thought, the significance of these views was hardly monumental. Then came the crisis of the seventeenth century and the early Saxon period, manifest, for example, in the isolation of Polish philosophy from what was going on in the world (counter examples, such as Kochański's correspondence with Leibniz, do not really change this picture). Also the Enlightenment, Romanticism and Positivism arrived to Poland some 30–50 years after they emerged in Western Europe. This state of affairs resulted in Polish philosophy being pluralist, but also eclectic, trying to accommodate many diverse threads and views.

LWS changed this situation. Twardowski's postulate that the world thought should be drawn on was quickly embraced. The Warsaw School of Logic is an excellent example. If in 1900 someone tried to foresee the development of logic after 1940, he would certainly not envisage Poland as the birthplace of a large school of logic. 'Wissenschaftliche Weltauffassung. Der Wiener Kreis', published in 1929, did not list a single Polish name. A year later (Neurath 1930), Warsaw logicians and philosophers are listed as representatives of scientific philosophy. Guests invited to the philosophical conference organised by the Vienna Circle before the International Philosophical Congress held in Prague in 1934 included a large group of Polish philosophers, mainly from the LWS. In the 1930 s, Poles, again mostly from the LWS, formed a significant group of lecturers. The influence of Polish logicians was huge, also in the philosophical dimension. Suffice it to cite the role of the semantic theory of truth (which changed the face of logical empiricism), the many-valued logic developed by Łukasiewicz, or the systems of Leśniewski. These facts contributed to the LWS being perceived as a logical school. There is no room here to analyse the significance of other ideas developed in the LWS. Let me limit my comments to saying that even if their international impact was limited, in the Polish philosophical community (at least partially) the conviction that Poles are not geese, but have a good philosophy of their own,[4] became well established. In particular, Polish analytical philosophers anticipated some of the important views which are today seen as classic, e.g., doubts about the dichotomy of analytic and synthetic statements (Ajdukiewicz, Tarski), methodological holism (Ajdukiewicz), semantic naturalism (Kotarbiński), the causal theory of time (Mehlberg), or the logic of induction (Hosiasson-Lindenbaumowa). I do not wish to claim that philosophy represented by the LWS was the only valuable philosophy in Poland, but the third of the School's accomplishments is the introduction of Polish philosophical thought and its achievements into the international community.

[4]This is a paraphrase of a famous quote from Mikołaj Rej: 'Among other nations let it always be known / That the Poles are not geese, have a tongue of their own.'

References

Dąmbska, I. (1948). Pięćdziesiąt lat filozofii we Lwowie. *Przegląd Filozoficzny, XLIV*, 14–25.

Ingarden, R. (1919). Dążenia fenomenologów. *Przegląd Filozoficzny, XXIII*(3), 118–156; reprinted in: Ingarden, R. (1962). *Z badań nad filozofią współczesną*. Warszawa: PWN.

Ivanyk, S. (2014). *Filozofowie ukraińscy w Szkole Lwowsko-Warszawskiej*. Warszawa: Semper.

Jadacki, J. J., & Paśniczek, J. (2006). *The Lvov-Warsaw School—The new generation*. Amsterdam: Rodopi.

Neurath, O. (1930). Historische Anmerkungen. *Erkenntnis, I*, 311–314.

Perreiah, A. R. (1991). Logic and philosophy in the Lvov-Warsaw School (Review). *Journal of the History of Philosophy, 29*, 149–150.

Struve, H. (1911). Słówko o filozofii narodowej polskiej. *Ruch Filozoficzny, I*, 1–11.

Tatarkiewicz, W., & Tatarkiewicz, T. (1979). *Wspomnienia*. Warszawa: Państwowy Instytut Wydawniczy.

Twardowski, K. (1911). Jeszcze słówko o polskiej filozofii narodowej. *Ruch Filozoficzny, I*, 113–115.

Twardowski, K. (1918). O potrzebach filozofii polskiej. *Nauka Polska, I*, 453–486.

Witwicki, W. (1920). Kazimierz Twardowski. *Przegląd Filozoficzny, XXIII*, IX–XIX.

Woleński, J. (1985). *Filozoficzna szkoła lwowsko-warszawska*. Warszawa: PWN.

Woleński, J. (1989). *Logic and Philosophy in the Lvov-Warsaw School*. Dordrecht: Kluwer.

Woleński, J. (1997). *Szkoła lwowsko-warszawska w polemikach*. Warszawa: Scholar.

Zegzuła-Nowak, J. (2017). *Polemiki filozoficzne Henryka Elzenberga ze szkołą Lwowsko-warszawską*. Kraków: Scriptum.

3

The Victims and the Survivors: The Lvov-Warsaw School and the Holocaust

Elżbieta Pakszys

1 Introduction

Human death, passing away, crossing over into a state contradictory to life, seems to be neither anyone's fault nor merit, since it is inherent to the nature of life, which begins, lasts, and ends. However, death can be caused by unavoidable internal conditions (diseases) or external circumstances (accidents or disasters); it can also be caused by others, intentionally or not. Natural death may be expected, quiet, and accepted. On the other hand, 'unnatural' death is sudden and unexpected, violent, disastrous, and often atrocious.

Dying is usually a very personal, private matter to human beings, especially when perceived consciously, even when it happens to many at the same times, for instance during a natural disaster or war.

In 1888, the German composer Richard Strauss (1864–1949) wrote the so-called tone, or symphonic, poem 'Tod und Verklärung' ('Death

E. Pakszys (✉)
Adam Mickiewicz University, Poznań, Poland
e-mail: pakszyse@amu.edu.pl

A. Drabarek et al. (eds.), *Interdisciplinary Investigations into the Lvov-Warsaw School*, History of Analytic Philosophy,
https://doi.org/10.1007/978-3-030-24486-6_3

and Transfiguration') to the poetry of his older friend, Alexander Ritter (1833–1896). It was dedicated to their late friend, Friedrich Rosch, who died in 1889. In it, the process of dying is depicted in four movements: Part I. *Largo*: The dying man is tormented by nightmares in his sleep. Part II. *Allegro molto agitato*: His suffering increases; memories of his life. Part III. *Meno mosso*: The dying man's life passes before him. Part IV. *Moderato*: His soul is liberated from his dead body and passes into the sphere of death. The movements represent consecutive stages of an ill person passing away, reconciled with his unavoidable death (cf. Golianek 2017).

Listening to this piece of music we think about death. Specifically, we think about the death of philosophers and about 'philosophical' death, and then ask ourselves: how did philosophers die? We might think of the lofty, sublimed, ritual suicide of Socrates, pictured in Plato's dialogues 'Apology' and 'Phaedo'; the quiet, rational, and calculated passing away of David Hume (anticipated in his autobiography); the untimely death of René Descartes in Stockholm, Sweden; and the equally dramatic, albeit in entirely different circumstances, death of Voltaire…

Perhaps, beyond the fact that they have been noticed and portrayed in better-known records, these cases of philosophers' deaths are not at all different from any other, ordinary human death. Unless it is an unexpected, forced 'inhuman' death. Such death was prevalent and happened on a mass-scale in the twentieth century.

The picture of death presented above, so 'aesthetically' imagined and typical of German post-Romantics, was shattered to pieces at time of the Holocaust, as concluded by Jean Amery (1977/2007) in an analysis of his own situation as a helpless intellectual (a philosopher) and an intellectually helpless prisoner of a Nazi German concentration camp. Although he survived Auschwitz, he had been robbed of any faith in the meaning of human existence in the world, where death equalizes the horror of dying, becomes a banal, mechanical, everyday fact, not even reflected upon. This, even after all these years, attests to the enormity of the perpetrated crimes and the still overwhelming powerlessness of the enslaved. Primo Levi attests to it as well; however, his stand is in direct opposition to Amery's overwhelming pessimism or negativism about the possible victory of human values (Levi 1958/2008).

Reflecting on the phenomenon of death in a gulag, so similar to that in Auschwitz, Gustaw Herling-Grudziński points to three main aspects of its horror: its suddenness, commonness, and anonymity (1946/1999: 196–198).

Relying on the testimony of witnesses and survivors, we will attempt to discuss this distressing subject of death in general as well as in particular as it concerns philosophers of the Lvov-Warsaw School during the Holocaust, perpetrated by the Nazi Germans during World War II.

Polish analytical philosophers were a quite randomly assembled group, not necessarily representative in sociological terms. Perhaps also imperfect for statistical purposes, since the group was not numerous enough. However, they can be perceived from a broader perspective as one or two generations of scholars, academics and intellectuals in the period between two world wars, who lived in the sphere of educational activities and philosophical influences of Kazimierz Twardowski and his students.

2 The Victims

In her brief account of war casualties among Polish philosophers, Daniela Tenner-Gromska (1948) describes the fate of about 60 individuals connected with this field of study in pre-war Poland, whose deaths appeared to be most often the result of World War II.

A comparison of her findings, covering only the period until the end of October 1946, will help us better understand the circumstances of these deaths and their nature—mostly violent, since connected, directly or indirectly, with the war. Some of them still remain unknown or cannot be confirmed as regards the date and other particulars, however.

From the total number of 60, five persons whose deaths are dated to the year 1938 or 1939, just before and after the outbreak of war (September 1, 1939), should be eliminated as not being its direct result.

Among the remaining 55 Polish philosophers whose lives were claimed by World War II, only a few died 'in the line of duty' or fighting in the 'field of glory', which in Jean Amery's terms would be described as a 'soldier's, heroic death'. They included: Jerzy Siwicki

(–1939), Michał Wasilewski (1906–1939), Jan Gralewski (1912–1943), an underground courier who was killed while trying to escape.

The group included six women (about 10%). One of them, Józefa Kodis (1865–1940), was not connected with the Lvov-Warsaw School, but the others were its important young representatives just before the war: Janina Hosiasson-Lindenbaum (1899–1942), Estera Markin (1903–1942), Eugenia Ginsberg-Blaustein (1905–1944), as well as Sabina Epstein-Weinberg (–1942) and Julia Dickstein-Wieleżyńska (1881–1943) who represented a different philosophical orientation. All of them suffered tragic deaths because of their Jewish origin.[1]

Nine or ten (about 20%) philosophers of the Lvov-Warsaw School died in Gestapo prisons or were killed during investigations conducted by the SS: Antoni Pański (1895–1942/1943) and Janina Hosiasson-Lindenbaum (–1942/1943) perished in the Vilnius prison.

There were five (10%) ghetto casualties: in Warsaw—Adolf Lindenbaum (–1941), Jakub Rajgrodzki (–1941) and Estera Markin (–1942), who was eventually transported to the death camp in Treblinka; Leopold Blaustein (1905–1944) and Eugenia Ginsberg-Blaustein lost their lives in the Lvov ghetto.

Another ten (about 20%) were victims of death camps. Scholars of the Jagiellonian University in Cracow (not connected with the Lvov-Warsaw School) were interned by the Nazi occupiers on 6 November 1939, and transported to the concentration camps in Sachsenhausen and Dachau. They were mostly elderly professors emeriti: Tadeusz Grabowski (1869–1940) perished in Sachsenhausen; Stefan Kołaczkowski (1897–1940) died after returning to Cracow; Joachim Matellmann (1889–1942), initially imprisoned in Sachsenhausen, died (probably) in the concentration camp in Mauthausen; Ignacy Chrzanowski (1866–1940) died in Sachsenhausen (Urbańczyk 1946/2014).

---

[1]Just as Edith Stein/St. Teresa Benedicta of the Cross (1891–1942), a German-Jewish philosopher, convert to Christianity who became a Carmelite nun and lost her life in the Auschwitz concentration camp, and Marthe Mourbel, a French-Jewish professor of philosophy at the high school in Angers, prisoner of the Ravensbrück concentration camp, who died during the camp's evacuation in 1945 (Helm 2015/2017: 913).

Edmund Romahn (1900–1943) lost his life in the death camp in Majdanek, Marceli Handelsman (1882–1945) in the concentration camp in Nordhausen/Erlagen; both had been associated with the Lvov-Warsaw School. Those who perished in the Auschwitz concentration camp were: Zygmunt Łempicki (–1943), Aleksander Kierski (–1943), Adam Stawarski (–1943), Jan Morsdorf (–1944).

To our knowledge, there were four individuals who committed suicide at the outbreak of the war: Stanisław Ignacy Witkiewicz (aka Witkacy 1885–1939), Salomon Ingel (1889–1942), Alfons Baron (–1940), Władysław Spasowski (1877–1941).

Five members of the group (about 10%) perished in the Warsaw Uprising (August 1944): one of the Uprising's chaplains, Rev. Jan Salamucha (1909–1944), Mieczysław (Michał) Treter (1883–1944), wounded in action—Karol Irzykowski (1873–1944) and Mieczysław Milbrandt (1915–1944), shot dead by the occupiers—Jan Łempicki (–1944).

3 The Survivors

Reports and research conducted after the war revealed cases of individuals who had been rescued from prisons and concentration camps, and also those who had survived the Holocaust in hiding. They included one of Kazimierz Twardowski's early students—Irena Panenkowa, née Jawicówna (1879–1968), who, like Daniela Tenner-Gromska (1889–1973), belonged to the first generation of women in the Lvov-Warsaw School.

Our own research (Pakszys 1997, 1998) revealed two outstanding Polish philosophers from the second generation of women in the Lvov-Warsaw School among the survivors: Dina Sztejnbarg, who assumed the war name Janina Kamińska, and later—after marrying Tadeusz Kotarbiński—was known as Janina Kotarbińska (1901–1997), and Seweryna Łuszczewska-Romahn (1904–1978). All of them survived concentration camps and resumed active academic lives after the war. They all shared one important experience, that of being a prisoner in the female concentration camp at Ravensbrück. However, not all of their names can be found in the camp register held by the German Nazis, later published in print (Kiedrzyńska 1965).

The Ravensbrück concentration camp was destined exclusively for female inmates; it was built by them and was operated during all six years of World War II (1939–1945). Out of the 132,000 women of different nationalities and ethnic backgrounds who were brought there from all over the Nazi-occupied Europe, about 92,000 lost their lives (Kiedrzyńska 1976). Polish women, also those of Jewish origin, constituted one of the most numerous national groups (totalling about 23,000). Looking for the names of the female philosophers mentioned above in the published lists of prisoners transported to the camp, we came upon the names of some interesting individuals who became well-known after the war.

Countess Karolina Lanckorońska (1898–2002), a respected art historian already before the war (since 1935, a docent at the Jan Kazimierz University in Lvov), was brought to Ravensbrück in 1943 (prisoner number 16076) for her participation in the underground resistance movement and for identifying those responsible for the murder of Lvov university professors.[2] She was released from Ravensbrück in April 1945 thanks to the Red Cross (Lanckorońska 2001).

Zofia Ryśiówna (1920–2003) (an outstanding actress after the war) was assigned her concentration camp number 7286 when she was brought to Ravensbrück on 12 September 1941. She was brought in a transport from Tarnów and imprisoned as a punishment for playing a key role in an action organized by the resistance movement to rescue an important underground courier, Jan Karski, from the Gestapo prison in Nowy Sącz (Karski 1944/1999).[3]

Considered an eminent painter and illustrator after the war, Maja Berezowska (1893/1898–1978) arrived at Ravensbrück in a transport from Warsaw and Lublin on 31 May 1942 and was assigned concentration camp number 11197. The offence against Hitler and the Third Reich which resulted in her imprisonment was publication of the Führer's caricatures.

---

[2]23 individuals perished in July 1941, including Tadeusz Boy-Żeleński (1874–1941).

[3]Describing this very action, particularly in chapters XII–XX of his book, at the end of p. 187 Karski makes a single remark: 'a woman was arrested', thus presenting another perspective on that action (cf. Ryś 2013: 16–22).

Both the draughtswoman and the future actress were active in what could be called the cultural life of Polish women prisoners' community within the camp. Maja Berezowska drew portraits of her fellow inmates—which was punishable by death—in order to document their personal identity and to beautify the gloomy reality of everyday life in the camp. Zofia Rysiówna took part in drama performances and recited poetry smuggled into the camp. The range and diversity of the secret cultural life of the different ethnic groups of women imprisoned in KL Ravensbrück, as described by Wanda Kiedrzyńska (1965) and Sarah Helm (2017), was truly astounding.

Irena Pannenkowa, mentioned above, also took part in the forbidden cultural life of Polish women in Ravensbrück, organizing philosophical talks popular among her female inmates; she delivered five talks on Socrates, for instance. She had studied at the Lvov University (1898–1905) and had been an early participant in the seminars lead by Kazimierz Twardowski, who had also been the reviewer of her doctoral thesis.

She had become a political activist before the war, writing a book which was her contribution to the discussion about the role of Józef Piłsudski in Poland's regaining independence.[4] She was also sent to Ravensbrück probably because of her involvement in the underground movement.

Unfortunately, our research concerning two other female philosophers of the Lvov-Warsaw School: Janina Kamińska/Kotarbińska and Seweryna Łuszczewska-Romahn, did not yield positive results in the form of dates, prisoner numbers, etc., probably because of their late (1944) arrival to Ravensbrück and a short or even transitional stay there, and perhaps also because the transportation lists to KL Ravensbrück are still incomplete.

On the other hand, in spite of her relatively short stay in KL Ravensbrück, a German-Jewish prisoner Helga Erdmann, who survived the camp and became an outstanding writer after the war, made her horrifying testimony public in 'Hollentor' (Lundholm 2014).

---

[4]Her book entitled *The Legend of Piłsudski,* published in 1922 under the penname Jan Lipecki, was mainly a polemic with the legend and a critical account of the cult of the Marshall.

The story of Janina Kamińska/Kotarbińska could be reconstructed based on her account given to us during an interview (1993). During the interview, she reminisced about being apprehended by the Gestapo, who had discovered a cell of the underground education system in which she was involved, and then transported from a prison in Warsaw first to the concentration camp Auschwitz, and then to Ravensbrück. She stayed there for about three weeks, and was then sent to Ravensbrück's sub-camp Malchow (operated from 1943 to 1945), from which she was rescued by the Swedish Red Cross in an operation conducted by Folke Bernadotte in April 1945. Kamińska/Kotarbińska made it safely to Malmö in Sweden (Bernadotte 1946; Helm 2017).

Seweryna Łuszczewska-Romahn was expelled from the Lvov University in 1939, and then arrested by the Gestapo together with her husband for their participation in the underground resistance movement, and probably also because of her 'foreign' origin. They were both sentenced to the death camp in Majdanek. She became a widow there (1943). She was first relocated to Ravensbrück, and later to the concentration camp in Buchenwald. At the end of the war, after her escape from the so-called 'March to Nowhere' (evacuation of camp prisoners), she was rescued by the US Army and worked as an interpreter for the UNNRA. She returned to Poland and was employed at the Poznań University in 1946. In 1969, she was appointed Chair of the Department of Logic, a position previously held by Kazimierz Ajdukiewicz (the Department later developed into the Institute of Philosophy at Adam Mickiewicz University). She retired in 1974 (Batóg 1979).

Nevertheless, we were able to investigate two cases of Holocaust survivors who had been only loosely associated with the Lvov-Warsaw School as philosophers, mostly through their polemical attitude, since they represented positions completely different from those of the School's representatives (Pakszys 2001; Zegzuła-Nowak 2017).

The case of Henryk Elzenberg (1887–1967) was brought to our attention by Professor Jan Tyburski at a conference in Warsaw in 2017. Elzenberg had survived the war in Vilnius, luckily avoiding deportation to the ghetto; in 1944, he was employed as a night watch in the carpentry workshop on a small property of Tadeusz Czeżowski (Elzenberg 1998).

Ludwik Fleck's (1896–1961) fate was more complicated. Before the war, he was an outstanding microbiologist-bacteriologist rather than a philosopher. He survived the Lvov ghetto thanks to the protection of Rudolf Weigl, who employed him at the Institute for Research on Typhoid Fever (Weigl, a Polish biologist, managed to save many other Polish Jews in the same way). Later Fleck was sent to concentration camps in Auschwitz and Buchenwald, where he survived, thanks to being useful to German Nazis in their efforts to contain outbreaks of typhoid fever within the concentration camps and beyond. After years spent in oblivion, he was recognized for his book, published in 1935, entitled 'Entstehung und Entwicklung einer wissenschaftlichen Tatsache: Einfuhrung in die Lehre vom Denkstil und Denkkollektiv' ('Genesis and Development of a Scientific Fact') as an outstanding forerunner of the sociological current within the philosophy of science, also in Poland (Fleck 1979/1986).

4 Conclusion

How did the Holocaust influence the Survivors, including the philosophers presented above, in their everyday life after the war? There is a growing body of literature commemorating this problem in both a general and a particular way, often revealing the darker side of Polish-Jewish relationships before, during, and after World War II (Jan Karski, Zofia Posmysz 1970/2017, Irena Sendlerowa, Nechama Tec, and many others).

Let us listen to the voice of one of the victims, a young Polish Jew, Itka Frajman Zygmuntowicz, who—many years later—explicitly described her state of utter deprivation when leaving the concentration camp. Her story is similar, at least to a certain degree, to those of Dina Sztejnbarg/Janina Kamińska/Kotarbińska and Seweryna Łuszczewska-Romahn.

> All on earth that I loved and held sacred I lost in the Holocaust, including nearly six precious years of my life. All on earth that I had left after liberation from Malchow, Germany, was my skeletal body, minus all my hair, minus my monthly cycle, a tattered concentration camp shift dress without undergarment, a pair of beaten up unmatched wooden

> clogs, plus my "badge of honour" - the large blue number 25673 that the Nazis tattooed on my left forearm on the day of my initiation to the Auschwitz inferno. I was homeless, stateless, penniless, jobless, orphaned, and bereaved… I had no marketable skills… Jewish homes, Jewish families, and Jewish communities were destroyed. I was a displaced person, a stranger, alive, but with no home to live in. I had no one to love me, to miss me, to comfort me, or to guide me. (Horowitz 1998: 370)

In regard to the Polish female philosophers whose survival stories are presented here, their war experiences were known best only to their closest family members and perhaps also some trusted, devoted students.

Janina Kamińska/Kotarbińska, whose mother survived outside of the Warsaw Ghetto in Poland, had someone to go back to after the war. She was able to achieve relative stability in life at her husband's side. However, was not her occasional coldness and acrimony, recalled by some of her students, caused by her tragic experiences of war? It is hard to say. In the interview she gave us, she never complained about any kind of harassment experienced from the Poles, especially at the Academy.

Tadeusz Batóg (1979), a student of Seweryna Łuszczewska-Romahn, mentioned her quite limited activity after the war, caused mostly by significant health problems. She supervised only three doctoral theses, and her scholarly publications after the war were not very numerous.

Did they leave any reminiscences of war perhaps, or personal diaries documenting those times? Nobody knows.

References

Amery, J. (1977/2007). *Jenseits von Schuld und Sühne. Bevaltigungsversuche eines Uberwaltigten/Poza winą i karą. Próby przełamania podjęte przez złamanego*. Stuttgart and Kraków: Wyd. Domini.

Batóg, T. (1979). Seweryna Łuszczewska-Romahn. *Studia filozoficzne, 1*, 189–194.

Bernadotte, F. (1946). *Koniec Trzeciej Rzeszy*. Katowice: Wyd. Awir.

Elzenberg, H. (1998). *Człowiek wobec wartości. Filozofia Henryka Elzenberga*. Warszawa: Wyd. Starkoz.

Fleck, L. (1979/1986). *Genesis and development of a scientific fact*. Chicago and London: The University of Chicago Press.

Golianek, R. D. (2017, April 21). *Gwiazdy światowych estrad; Jereme Rhorer, dyrygent, Martin Stadfeld, fortepian, Program koncertu Filharmonii Poznańskiej*. Poznań.

Gromska, D. (1948). Philosophes polonaise mortes entre 1938 et 1945. *Studia Philosophica, III*, 31–97.

Helm, S. (2015/2017). *If this is a woman. Inside Ravensbrück: Hitler's Concentration Camp for Women/Kobiety z Ravensbrück; Życie i śmierć w hitlerowskim obozie koncentracyjnym dla kobiet*. Warszawa: Wyd. Prószyński i S-ka.

Herling-Grudzński, G. (1999). *Inny świat. Zapiski sowieckie: Krzyki nocne*. Warszawa: Wyd. Czytelnik.

Horowitz, S. H. (1998). Women in Holocaust literature: Engendering trauma memory. In D. Ofer & L. J. Weitzman (Eds.), *Women in the Holocaust*. New Haven and London: Yale University Press.

Karski, J. (1944/1999). *Story of a Sacred State/Tajne państwo*. Warszawa: Wyd. Twój Styl.

Kiedrzyńska, W. (1965). *Ravensbrück kobiecy obóz koncentracyjny*. Warszawa: Wyd. Książka i Wiedza.

Kiedrzyńska, W. (1976). *By nie odeszły w mrok zapomnienia*. Warszawa: Udział kobiet polskich w II wojnie światowej.

Lanckorońska, K. (2001). *Wspomnienia wojenne 22.IX.1939–5.IV.1945. Ravensbrück*. Kraków: Wyd. Znak.

Levi, P. (1958/2008). *Se questo e unuomo/Czy to jest człowiek*. Kraków: Wyd. Literackie.

Lundholm, A. (1988/2014). *Hollentor/Wrota piekieł Ravensbrück*. Warszawa: Wyd. Karta.

Pakszys, E. (1997). Kobiety w filozofii polskiej. Dwa pokolenia Szkoły Lwowsko-Warszawskiej. In E. Pakszys, D. Sobczyńska (Eds.), *Humanistyka i Płeć II. Kobiety w poznaniu naukowym wczoraj i dziś* (pp. 173–191). Poznań: Wyd. Naukowe UAM.

Pakszys, E. (1998). Women's contribution to the achievements of the Lvov-Warsaw School: A survey. In K. Kijania-Placek & J. Woleński (Eds.), *The Lvov-Warsaw School and contemporary philosophy* (pp. 55–71). Dordrecht: Kluwer Academic Press.

Pakszys, E. (2001). Indywiduum wobec kolektywu myślowego Ludwika Flecka. Pozycja kobiet w filozofii analitycznej. *Zagadnienia Naukoznawstwa, 1*(147), 135–146.

Pannenkowa Irena/Jan Lipecki. (1923). *Legenda Piłsudskiego*. Poznań: Wlkp. Księgarnia Naukowa.

Posmysz, Z. (1970/2017). *Wakacje nad Adriatykiem*. Kraków: Wyd. Znak.

Posmysz, Z. (2017). *Królestwo za mgłą. Z autorką Pasażerki rozmawia Michał Wójcik*. Kraków: Wyd. Znak.

Ryś, Z. (2013). *Wspomnienia kuriera: Akcja „S"*. Nowy Sącz: Polskie Towarzystwo Historyczne.

Urbańczyk, S. (1946/2014). *Uniwersytet za kolczastym drutem. Sachsenhausen i Dachau*. Kraków: Wyd. Literackie.

Zegzuła-Nowak, J. (2017). *Polemiki filozoficzne Henryka Elzenberga ze Szkołą Lwowsko-Warszawską*. Kraków: Wyd. Scriptum.

4

The Lvov-Warsaw School on the University and Its Tasks

Włodzimierz Tyburski

1 Introduction

By way of an introduction to this chapter, it is perhaps worthwhile noting that the reflection on the tasks and role of the university, the ethics of scholars, and academic education in Poland has a long tradition.

Already the first professors of the Krakow Academy, including Matthew of Krakow, Bartholomew of Jasło, Stanisław of Skarbimierz, Paul of Worczyn, Benedykt Hesse of Krakow, Jan of Ludzisko considered this problem and emphasised its importance. They pointed out and demonstrated that 'the state reaches an equilibrium and grows in salvific power' when its rulers 'found universities and recruit the help of scholars in order to effectively eliminate faults and shortcomings afflicting their patrimony' (Domański et al. 1989: 41).

W. Tyburski (✉)
Institute of Philosophy and Sociology,
The Maria Grzegorzewska University, Warsaw, Poland
e-mail: tyburski@umk.pl

A. Drabarek et al. (eds.), *Interdisciplinary Investigations into the Lvov-Warsaw School*, History of Analytic Philosophy,
https://doi.org/10.1007/978-3-030-24486-6_4

In one of his famous homilies, Stanisław of Skarbimierz (ca.1365–1431), the first Rector of the Krakow Academy, encourages his fellow countrymen to study at the Academy, where

> mannerly speech is taught by grammar, sophisticated talk by rhetoric, subtlety in reasoning by logic, the ability to calculate by arithmetic, proficiency in measurements by geometry, rhythms, harmony of tones and intervals by music, the ability to predict future events by theory, self-control by monasticism, household government by economics, government of large groups of people like societies by politics'. [...] [The Academy is where] 'problems are resolved, disputes are settled, the conscience is cleansed; it is the place where churches are governed in accordance with law and canons, where there is salvation, where there is life, where there is peace, where there is safety, where theology enjoys its glory and honour; where there is everything that is elsewhere sought for the sake of salvation'. (Domański et al. 1989: 82–85)

2 The Ethos of a University Professor

The first code of ethics for academic professors was created by Szymon Marycjusz of Pilzno (1516–1574). In his work entitled 'De scholis seu academiis' ('On Schools or Academies'), he presented a broad catalogue of intellectual and moral virtues describing an ideal professor of liberal arts. Aside from virtues inherent to the intellectual picture of an academic teacher, such as excellent mastery of the subject he is to teach, high scholarly proficiency, cognitive skills, and highly valued didactic abilities, he presents and describes moral virtues constituting the role model of an academic teacher and the norms of ethics in education. The principal ones he lists are diligence, justice, 'honesty and purity of mores'. He also emphasises the strict relationship between the ethical and the academic criterion. For one is not a good scholar unless 'he is also a good man' (Szymon Marycjusz 1925: 143). Only a professor who combines knowledge with honest life and a good character may 'draw the youth under his influence' (Szymon Marycjusz 1925: 144).

Andrzej Frycz Modrzewski (1503–1572) believed that school has an important role to play in ethical and civic education, as it creates the best conditions for the development of cognitive powers and the moral and civic improvement of man. He pointed out that all individual and social evil stems primarily 'from ignorance'. Ignorance is a kind of blindness to that which is wise, valuable, and precious. Enlightenment in the sphere of these values is the primary task of the academy. Therefore, the road to repairing man and the Republic leads through making education at all levels universally accessible, and through its reformation and improvement.

Along with Marycjuszof Pilzno, Szymon Stanisław Makowski (ca. 1612–1683) was one of the first scholars to develop a set of norms and obligations describing the ethos of a university professor and his relations with students. He created one of the first codes of academic ethics and deontology. The code begins with a reminder of the three fundamental obligations of a teacher: the first one concerns the quality of knowledge to be passed on; the second refers to the obligation to eliminate errors in the acquisition of knowledge; and the third is concerned with the teacher's commitments to the students' education, development of the right kind of intellectuality, and to their upbringing. These general obligations of a university professor are discussed in detail across a broad catalogue of individual duties. As a long-standing Rector of the Krakow Academy, he was engaged in a dispute with the Bishop of Krakow, Andrzej Trzebiecki (the Academy's Chancellor), most emphatically and persistently standing up for the Academy's autonomy.

Representatives of the Enlightenment: Stanisław Konarski, Antoni Wiśniewski, Stanisław Popławski, Hugo Kołłątaj, Stanisław Staszic, brothers Jan and Jędrzej Śniadecki spoke on many occasions on the ethics of scholarly work and academic teaching. Antoni Wiśniewski (1718–1774), professor of philosophy, physics, and mathematics at Collegium Nobilium, a leading representative of *philosophiae recentiorum*, wrote about the need for an ethics of learning and moral teaching. There can be no academic education, he claimed, without moral teaching, and warned that 'supreme learning without virtue may become not a contribution to but the doom of the human community' (Wiśniewski, 1760: 327).

Hugo Kołłątaj (1750–1812) was one of the main founders and activists of the Commission of National Education. On behalf of the Commission, he first reorganised the Nowodworski Gymnasium in Krakow, then implemented thorough reforms of the Krakow Academy where he held the office of Rector for a number of years. He introduced revolutionary changes in the curriculum, removing scholastic education and such anachronistic disciplines as astrology, for example. Instead, he introduced the study of the laws of nature and natural sciences taught in the spirit of philosophical empiricism, including studies in geology, whose development was strongly encouraged as a contribution to the mining industry. He was committed to the practical dimension of education at the Academy.

A similar role was played at the Academy of Vilnius by its Rector Marcin Poczobutt (1728–1810). Acting on a mandate from the Commission of National Education, he reformed the Academy, introduced natural and mathematic sciences as well as modern engineering studies, extended and equipped the astronomical observatory.

As regards the development and organisation of sciences, a particularly great contribution was made by Stanisław Staszic (1755–1828) through his involvement in the Warsaw Society of Friends of Learning. During the Duchy of Warsaw period, he became actively involved in the organisation of education. He continued his work in the Congress Poland, becoming particularly engaged in the development of scientific disciplines relevant for economy, industry, and technology. He raised funds for scientific research, particularly in the disciplines he was most interested in: geology, mining, and metallurgy.

Jan Śniadecki (1756–1830) held the office of Rector at the Vilnius Academy for eight years. He was committed especially to the development of colleges and lower ranking schools subordinated to the University. In addition, he was head of the astronomical observatory, holding the office until he retired in 1824.

Leading representatives of Polish positivism: Aleksander Świętochowski, Julian Ochorowicz, Feliks Bogacki, Adam Mahrburg, also commented on the ethics of scholarly work and academic education. This intellectual formation was involved in the work of laying foundations and propagating science and the ideology of learning. They were most ardently advocating objectivism in science, reliability in research, the need for scientific criticism. They were champions of science free from value judgements, and explicitly supported the attitude of an objective, 'unsentimental researcher'.

3 Kazimierz Twardowski and the Modern University

The organisation and tasks of the modern university were discussed in the academic circles with unprecedented intensity already at the turn of the twentieth century. The discussion was initiated by the founder and creator of the Lvov-Warsaw School, Kazimierz Twardowski. For nearly three decades, he set the tone of the great debate held in the Polish academic community about the contemporary role of the university and its scientific and academic tasks. Not only did he initiate discussions on the subject, but put their outcome into practice, along with his own ideas and thoughts as the organiser of academic life in Lvov and across the country. Scholars studying his legacy agree that it was not only in research, but also in his organising and teaching activities that Twardowski's achievements deserve the highest recognition. He made a great contribution to the work of organising and establishing Polish institutions of higher education and scientific societies, particularly through his active involvement in the Society of Academic Teachers.

A voluminous festschrift published by the Society, containing Twardowski's addresses and speeches delivered during his presidency of the Society, reveals how active, generous, and multidirectional his work was. In the years 1908/1909, he became involved in the work of reorganisation and modernisation of the chancellery of the Lvov University. He developed the first regulations for university authorities, administration, and students (Twardowski 1908). He devoted a lot of attention, as well as conceptual and organisational effort to these matters while holding the office of the University's Rector for three terms. Here is what his closest student, Tadeusz Czeżowski, whom Twardowski appointed head of the University's chancellery, writes in his autobiography about Twardowski's work as Rector:

> It was hard work. The Rector worked at the University from 8.00 am to 2.00 pm, and from 4.00 pm to 8.00 pm –ten hours a day. And he expected me to work the same hours. And yet, this cooperation brought me closer to the Professor, so much so that he became like a foster father to me after I lost my own father to a long and serious illness in 1915. At the same time, this work was a lesson in administration under the excellent guidance of Twardowski, which proved to be very valuable in my later life'. (Czeżowski 1977: 432)

Twardowski also held the office of Rector during World War I, a particularly difficult, dramatic period in the University's history. In the reborn Poland, he drafted a bill on academic institutions, and participated in the subsequent work of reforming the system of middle and high-level education. As regards organisational matters, 'he was regarded as an authority both in the Austrian Ministry of Education, the National Board of Schools in Galicia, and by each subsequent minister in charge of the science and education department in the free Poland' (Łempicki 1938: 39).

Twardowski was an ardent advocate of the independence of universities, freedom of scientific research, and the importance of science. He placed much emphasis on this issue in his excellent, often cited address 'O dostojeństwie Uniwersytetu' ('On the Dignity of University') delivered in 1932 on the occasion of his being awarded the honorific doctorate of the University of Poznań.

He said that the university

> [...] has every right to demand that its spiritual independence be respected by all, and to defend itself against any overt and covert attempts at submitting its scientific work to anyone's control or command. At the same time, the University has the duty to avoid anything on its part that might compromise that independence, or even give the impression of surrendering to any influences or tendencies which have nothing to do with scientific research and its purposes. It must ward off anything that does

> not serve the discovery of scientific truth; it must keep appropriate distance between itself and the current of common everyday life flowing past its walls, the turmoil of clashing social, economic, political or any other currents. In the midst of all of these struggles and skirmishes between all of these diverse streams and currents, the University must stand firm like a lighthouse which shows the way to ships sailing across rough seas, without ever dipping its light in those waves. For if it did so, the light would go out, and the ships would be left without their guiding star'. (Twardowski 1933: 6–7)

These words may also be read as a 'covert polemic with the contemporary threats to universities brought by the Jędrzejewicz reforms of 1932 and 1933, which limited the autonomy of universities' (cf. Mincer 2002: 104), and a warning against tendencies to politicise academic life which intensified in the 1930s. He warned politicians against attempting to subordinate universities or trying to control or check them in any way.

> Those who found and provide for universities would demonstrate utter ignorance of what the essence of University is if they tried to shackle it, rejecting some of the results of its work in advance, or suggesting what outcomes would be welcome. And even if the results of scientific work carried out at the University were unfavourable to those to whom it owes its existence, this should never legitimise imposing any reins or placing any constraints. For scientific research may only develop and bring fruit if it is not restricted or threatened in any way. (Twardowski 1933: 8)

4 Tadeusz Czeżowski on the Tasks of the University

Tadeusz Czeżowski, one of Twardowski's closest students, took over, continued and developed many of the ideas, concepts and views of his master, including those concerning the status, role and functioning of academic institutions, and the tasks and vocation of scholars. His faithfulness to Twardowski's teaching and directions was documented in the practical contribution he made to the organisation of academic life, as well as in many scientific publications where he discussed the functioning and organisation of institutions of higher education and academic teaching. He acquired practical experience first as head of the chancellery of the Lvov University, and then, in the free Poland, at the Department of Academic Institutions of the Ministry of Religious Affairs and Public Education (MWRiOP). Also his subsequent work at Stefan Batory University in Vilnius provided him with ample experience, which later brought the fruit of valuable publications. A particularly noteworthy project in this regard was the 'Collection of Acts and Regulations on University Studies and Other Provisions Relevant to University Students, Including in Particular the Stefan Batory University in Vilnius' he compiled (Czeżowski 1926: 374). The first part of this very practical and useful collection of regulations discussed the organisation of university studies; the second part focused on the organisation of studies at individual departments; and the third part dealt with scholarships, military service, railway discounts, etc. In the Introduction, Czeżowski wrote about the primary tasks of universities:

> They are in the service of science, while at the same time being the highest vocational schools for lawyers, doctors, teachers, etc. These tasks are not mutually contradictory, but supplement each other. Thorough preparation to practicing a profession needs to be based on scientific foundations, and the path to scientific knowledge can only be shown by one who works in the field. Therefore, an academic teacher can only be a scholar, but never a practitioner, however excellent. At the same time, the task of providing scientific education does not hinder, but in fact contributes to theoretical research conducted at universities; this way science remains in touch with real life and its problems, and from among the crowds now enrolling in university courses for the practical reason of obtaining academic education, individuals may be singled out who devote their efforts to science alone. For from the two ends served by the university, the first one, science, has primacy, and vocational education only comes second. (Czeżowski 1926: VII–IX)

Czeżowski, just like Twardowski, most emphatically advocated the independence of universities, freedom of scientific research, and their autonomy from the state and politics. In his lecture 'O stosunku nauki do państwa' ('On The State's Attitude to Science') delivered during a formal meeting of the Józef Mianowski Fund—a Foundation for the Promotion of Science on 12 May 1932, he listed a catalogue of basic principles on which appropriate relationships between the state and institutions of higher education and science should be founded. He pointed out that the mutual relationship between the state and science could not be seen merely as the government and state institutions financially supporting the development of science in order to reap the benefits of its findings. The modern state displayed a tendency to subordinate various areas of social life, including science, and thus to 'nationalise' them. The state wanted to be able to influence the development of science and to regulate it, both in terms of research and education, in line with its objectives (Czeżowski 1933: 6). In the area of their mutual relationship, fundamental discrepancies sometimes appeared, since the intentions, goals and, ideals followed by the state did not always coincide with the goals and ideals of science. Indeed, one could say that 'their ideals are usually different'. While 'the state's ideal is a certain *good*', the 'ideal of science is *truth*', and sometimes these two ideals may stand in opposition to each other depending on how this 'good' is understood (cf. Czeżowski 1933: 6).

Another danger appears when the state decides, for utilitarian reasons and for the sake of immediate benefits, that certain areas of scientific research are not particularly useful, and consequently aims to limit or even remove them from the academic curriculum, cutting down on subsidy which enables their development. Still another threat to science appears when the state decides, based no longer on financial, but on ideological reasons, that some areas of research are dangerous. Pushing its way forward towards truth, scientific thought often challenges dogmas, sanctified, seemingly inviolable convictions, and if they happen to be the foundation of the state's system, then the state acts as a censor or simply prohibits research in certain areas. In its search for truth, science does not shy away from criticism or revision of what has been achieved so far, and is against proclaiming any dogmas which are excluded from scientific discussion and criticism. The state may also embrace politics which are faulty, or even detrimental to science, for some other reasons. This happens when the ruling class is characterised by incompetence, inefficiency and indolence, or when current politics subordinate science to various interests. In such circumstances, the government collides with science and the way it should be properly pursued, which may result in lowering the standards of scientific research, and sometimes in a depravation of researchers, some of whom are willing to compromise scientific reliability for the sake of their career.

As a consequence of this collision between the state and science, Czeżowski says, not only does the condition of education and the level of scientific research deteriorate, but many other situations emerge which are undesirable, if not detrimental to the interests and goals of the society and the state. If the government restricts and hinders scientific freedom, it does so to the detriment of state and public weal. Science may, for example by criticising the state's systemic or social foundations, contribute significantly to a gradual evolution of its forms in the direction of progress and common good. After all, criticism of the existing system, particularly of its obsolete and ineffective forms, does not have to be subversive, but quite on the contrary—it may be 'the most effective way of safeguarding against such tendencies' (Czeżowski 1933: 6). Czeżowski asserts that scientific criticism may contribute to social stabilisation of the state, as by identifying in advance, in a scientifically documented way, from an objectivist standpoint, the sources of crises, 'it may blunt the edge of radical slogans by demonstrating their error' (Czeżowski 1933: 9), and thus hold back those who use catchy rallying cries in order to abolish or undermine the existing order. This way, scientific criticism focused on social problems often proves beneficial to public weal. Governments which do not understand this and which shackle the freedom of spirit, word, and criticism, in fact contribute to dissemination of mistaken ideas, more detrimental to the society and the state than the alleged losses caused by scientific criticism—and this may result in violent upheavals leading to decline and ignorance.

Many disruptive consequences may also arise from the fact, Czeżowski says, that brakes are put on those areas of scientific research which cannot demonstrate an immediate application of their results. Such practices are a threat to science itself. After all, it is 'one organic whole, it springs from one common source' and aims at 'constructing a rational picture of the whole of reality' (Czeżowski 1933: 9). Therefore, all spheres of reflection and 'scientific study are mutually dependent upon one another, so if one of them is neglected, this may have a negative effect on the other in most unexpected ways' (Czeżowski 1933: 8).

On the other hand, Czeżowski points out, we cannot foresee what practical applications the findings of theoretical sciences may have in the future. The history of science has proved more than once that studies into matters apparently very distant from practical applications have opened up entirely new areas essential for the development of technology, or proved to have unexpected benefits in political and social life. One can never say in advance that some studies will not prove useful in the future, so hampering some fields of research may be directly or indirectly linked to the loss of certain values, also in material terms.

Therefore, if those in power refuse to embrace a mistaken idea about the best interest of the state, but understand it correctly, then they should see science as a tool, a means for achieving it, for 'there is no material difference between the ideal of good and the ideal of truth' (Czeżowski 1933: 10). Clearly, science is a significant instrument in realising state interests and serving public weal, which does not disparage its value in any way. The point is to make appropriate use of this instrument, so that

> the state does not overexploit science but, like a good steward, provides it with all it needs to live on and thrive, without forcing it to perform in ways in which it cannot, just like one does not use a delicate precision knife to do the work of an axe or a saw. [...] The state will benefit from what science has to offer not by making it perform particular functions, but by recognising its independence and distinction, ensuring appropriate conditions for its development. This is when science will bring the most abundant harvest for the state of its own accord. (Czeżowski 1933: 11)

In nearly all of his comments on science, higher education and academic learning, Czeżowski, like Twardowski before him, is a most emphatic champion of the independence and full autonomy of the University and freedom of scientific research, arguing that science will not achieve its goals unless it is not made to 'serve any other ends', unless school authorities, teachers, and students place science in the foreground, and education is given priority 'to which other values are only supplemental' (Czeżowski 1933: 12).

Czeżowski often returned to the issue of academic institutions in other lectures and speeches he delivered. One of the most characteristic of his comments on the subject was presented a dozen years later in a book entitled 'O uniwersytecie i studiach uniwersyteckich' ('On the University and Academic Studies') (Czeżowski 1970). He reduces the multiplicity of tasks assigned to modern universities to four basic ones: (a) scientific research; (b) education of academics and researchers; (c) vocational training—teaching university graduates to combine high theoretical knowledge with its practical applications; and (d) popularisation of scientific findings. These tasks are to be reflected in the modern organisation of academic institutions. The basic principles of organising a university include: universality, self-government, freedom of teaching and learning.

Another significant contribution to the subject was his address delivered on 21 February 1970 in the lecture hall of Toruń University on the occasion of his 80th birthday. Czeżowski devoted this long lecture to issues which, as he emphasised, 'always mattered to him a lot'. It was, naturally, the issue of the role and activity of the university and academic education. Once again, he referred to the functioning and organisation of academic institutions, this time emphasising those transformations and evolutionary processes which universities underwent in the first half of the twentieth century. Remarking on changes in the organisation of academic institutions and university education, he put emphasis on those features of the university which were permanent and which enabled it to perform its tasks as a research and academic institution 'that provides training in so-called academic professions, i.e. those which require mastery of the scientific method' (Czeżowski 1970: 300).

The comments and remarks both Kazimierz Twardowski and Tadeusz Czeżowski made about the primary duties and tasks of the profession and ethos of a scholar are very similar as well. 'An academic,' Kazimierz Twardowski writes, 'is first of all a servant of objective truth, its ambassador and advocate to the youth and the society. It is a high and noble service, but one which requires not only appropriate intellectual qualifications and relevant expertise, but also great fortitude and strength of character'. To this description Czeżowski adds some more comments, stressing that an academic should be 'a creative researcher, making an independent, personal, valuable contribution to the progress of science through his scientific work, able to teach well, having an impeccable civic record, and characterised by high moral standards' (Czeżowski 1946: 18).

5 Conclusion

We have referred here mainly to the views and analyses of Kazimierz Twardowski and Tadeusz Czeżowski concerning the status, role, and significance of the university. In fact, however, they represent the standpoint of the entire Lvov-Warsaw School. All of its representatives shared the belief, expressed by the two scholars, that the university: (a) is supposed to serve the ideal of truth above all; (b) harmoniously combines both research and educational functions; (c) trains creative individuals characterised by an integrity of intellectual, moral, and aesthetic culture; (d) contributes to cultural development; and (e) influences its immediate and more distant surroundings through practical applications of its findings.

Let us notice that this concept of university differs from the proposals we hear today, aimed at restricting the broad range of its functions and tasks, and transforming it into an institution performing specific tasks, mainly those of a service provider. They are based on the mistaken—to my mind—belief that the university in its classic form no longer has any raison d'être in the modern society. Such a remodelled university would in fact be limited to the provision of educational services and create, or rather 'produce' knowledge and innovation corresponding to the current, changing demands of the market and the social and economic development.

The university concept proposed by the Lvov-Warsaw School and the contemporary proposals for an academic institution conceived as a facility providing educational and research services, determined by their immediate usefulness, are two very different institutions. They are different in their structure and functions, they have different goals and follow different standards. For example, the community of students and teachers cultivated in the classical university, the *universitas magistrorum et studiorum*, is nearly non-existent in the new model. Also the culture-making function of academic institutions, once so important, is now disappearing. These adverse phenomena are the result of more extensive transformations we are witnessing in our times. 'The modern man', Czeżowski points out, 'pursues record-breaking, peak achievements at the cost of limiting himself to narrow specialisation. This race is being run in every field – which leads to enormous conflicts in the social and politic sphere' (Czeżowski 1963: 165). The Professor believes that we should turn back on the path we have been following, 'away from the ideal of records, and back to the classic ideal of moderation and harmony' (Czeżowski 1963: 166).

We should note that an academic institution which provides narrowly understood educational and research services has the following characteristics: (a) interpersonal relationships based on the community of students and teachers, so important in the traditional university, are now being replaced by the relationship between clients and service providers; (b) the ideal of 'creative individuality, characterised by an integrity of intellectual, moral and aesthetic culture' is replaced by the figure of a specialist equipped with the necessary quantum of knowledge required to perform a profession; (c) the pedagogical function is largely marginalised; (d) the research function is determined by the principle of immediate usefulness and profitability; (e) basic research is pushed to the background; (f) the 'freedom of curriculum' is severely compromised for the sake of rigidly and arbitrarily constructed programmes; and (g) the primary measure of success for new institutions is the mass demand for their services.

It is therefore a different concept of an academic institution, very distant from the classical ideal of university. It cannot replace the concept of university lined out by Twardowski and Czeżowski. Abandoning the classical ideal and transforming the university into an institution which provides educational services and responds to changing demands of the market in fact means giving up on the universal goals pursued by the university performing its cognitive, educational, and pedagogical functions, engaged in forging creative individualities—as Twardowski and Czeżowski would have it.

These universal tasks cannot be performed by an academy understood as an institution focused on immediate usefulness, of a limited potential that quickly runs out, immersed in a particular reality and together with it subject to its fluctuations and transformations. It may certainly fulfil its immediate tasks by situating itself next to the university in its classical meaning, but not by replacing it or causing its actual decline. That is why the autonomy of university and independence of scientific research—strongly emphasised by both scholars—is of such fundamental importance. This means that the scientific point of view must not be displaced by political, particular, partisan, narrowly utilitarian reasons; otherwise, the academy would forfeit the most important purpose of its existence. In the modern society, the university as an institution with a long-term educational and cognitive mission is just as indispensable now as it was in the times when Twardowski and Czeżowski sketched out its structural and functional profile. And it is at such a university that men of science may implement the model of scientific work proposed by these two thinkers, and fulfil the obligations arising from the scholarly ethos they introduced. In such a model of university education, the academy will be able to most effectively fulfil its obligations by teaching and developing the creative personality of its students and alumni.

References

Czeżowski, T. (1926). *Zbiór zasad i rozporządzeń o studiach uniwersyteckich oraz innych przepisów ważnych dla studentów uniwersytetu ze szczególnym uwzględnieniem Uniwersytetu Stefana Batorego w Wilnie*. Wilno.

Czeżowski, T. (1933). *O stosunku nauki do państwa*. Wilno.

Czeżowski, T. (1946). *O uniwersytecie i studiach uniwersyteckich*. Toruń.

Czeżowski, T. (1963). Trzy postawy wobec świata. *Odczyty filozoficzne* Toruń.

Czeżowski, T. (1970). Przemówienie na uroczystości dnia 21 marca 1970 w auli Uniwersytetu Toruńskiego. *Studia Filozoficzne, 3*.

Czeżowski, T. (1977). *Wspomnienia. (Zapiski do autobiografii)*. Kwartalnik Historii Nauki i Techniki, XXII.

Domański, J., Ogonowski, Z., & Szczucki, L. (1989). *Zarys dziejów filozofii w Polsce. Wiek XIII–XVII*. Warszawa: PWN.

Łempicki, S. (1938). Rola Kazimierza Twardowskiego w uniwersytecie i społeczeństwie. In *Kazimierz Twardowski. Nauczyciel – uczony – obywatel. Przemówienie wygłoszone na akademii żałobnej urządzonej w auli Uniwersytetu Jana Kazimierza w dniu 30 IV 1938 przez Senat Akademicki, Radę Wydziału Humanistycznego Uniwersytetu Jana Kazimierza i Polskie Towarzystwo Filozoficzne*. Lwów.

Mincer, W. (2002). Uniwersytet Tadeusza Czeżowskiego. In J. Pawlak (Ed.), *Filozofia na uniwersytecie Stefana Batorego*. Toruń: Wyd. UMK.

Szymon Marycjusz z Pilzna. (1925). *O szkołach czyli akademiach*. Kraków.

Twardowski, K. (1908). W sprawie Uniwersytetu Lwowskiego. *Muzeum, II*, 2.

Twardowski, K. (1933). *O dostojeństwie uniwersytetu*. Poznań.

Wiśniewski, A. (1760). *Rozmowy w ciekawych i potrzebnych, filozoficznych i politycznych materiach w Collegium Nobilium Warszawskim Scolarum Piarum miane*. Warszawa.

5

The Axiology Project of the Lvov-Warsaw School

Anna Drabarek

1 Introduction

The very essence of being human entails taking action, and in the imperative to act a special place belongs to values which call for or prohibit particular behaviours. Obligation, which in this context becomes an indispensable category, cannot be determined theoretically, however; it is specific to a particular person and to a particular behaviour, taking place in unique circumstances.

Philosophers of the Lvov-Warsaw School, while not explicitly setting out to develop an axiology, did contribute, however, to creating an axiological project with their discussion of values, the ways of discovering and implementing them; their project is still discussed with great interest and provides inspiration for contemporary axiology.

A. Drabarek (✉)
Institute of Philosophy and Sociology,
The Maria Grzegorzewska University, Warsaw, Poland
e-mail: adrabarek@aps.edu.pl

A. Drabarek et al. (eds.), *Interdisciplinary Investigations into the Lvov-Warsaw School*, History of Analytic Philosophy,
https://doi.org/10.1007/978-3-030-24486-6_5

This axiological project is distinguished from many others by the clarity of thought, commitment to scientific precision and independence of reflection on moral values, as well as accuracy of moral arguments. It is a realistic axiology, based on moderation; it does not emphasise or privilege any particular way of presenting or judging values. In this approach, ethical codes give way to a realistic attitude to reality, so that actual situations and specific circumstances considerably affect the standards of conduct. In the opinion of Lvov-Warsaw School ethicists (which may and has indeed given rise to numerous controversies), rightful acts being a consequence of moral choices are not and cannot be determined by a worldview or by any metaphysical or scientific considerations.

The School's axiological reflection certainly emphasised the role of man, endowing him with moral autonomy in creating rules of conduct, as long as the fundamental condition was met of such rules being the only possible ones in the circumstances concerned. Therefore, for each situation in life there is only one right thing to do, and that is what should be done. The axiological project presented by Kazimierz Twardowski and his followers is based on a priori and intuitive assumptions, because good is considered to be an absolute value. This type of reflection on values can be referred to as empirical axiology, as it talks both about objective as well as individual and relative rules of moral conduct.

As the study of values, considering the problem of good and the related notions of act and happiness, axiology is based on four types of reasoning which generate attitudes to the mutual relationships between truth, act and morality. The first attitude is referred to as that of a prophet, as it foretells a time in which truth, act and good will come to exist as one. The second, referred to as that of a sage, is based on wisdom and experience, and looks for the conditions of unity and concomitance of these factors. In the third, technical attitude, each of these elements is considered and analysed, and their presumed relationship is often put in parentheses. Finally, the fourth attitude, referred to as parrhesiastic, consists in understanding truth, act and morality so that none of these three elements may exist without the other two. The attitude of a parrhesiastes is often identified with uncompromisingness, frankness and nonconformism. When we ask about happiness as the most keenly desired goal and meaning of life, one which is always attractive irrespective of where or when the question is asked, we discover that these four attitudes: that of a prophet, sage, technician and parrhesiastes, are still encountered in contemporary reflections on the problem.

In the views of Lvov-Warsaw School philosophers who deliberated upon the problems of truth, act, morality and the resulting category of happiness, we will not find guidelines provided by prophets or sages, but we can certainly study their technical and methodological reflections; after all, they were masters of the analytical method. We may also consider the parrhesiastes' attitude, proclaiming the truth about man and his moral choices without regard to conditions or circumstances in which it is uttered.

2 Objectivism in the Understanding of Good

While Władysław Tatarkiewicz (1989: 213) was not a student of Twardowski, founder of the Lvov-Warsaw School, his works were nevertheless written in accordance with the School's thematic and methodological standards. In his autobiographic article, he said that Aristotle was to him the most reasonable of all philosophers, and added that Aristotle was more cautious than any other classics in proclaiming truths about happiness. His caution was exemplified in the golden middle rule. Tatarkiewicz believed Aristotle to be the master of common sense, and called his ethical views ethics of moderation and amity. It can be conjectured that the concept of happiness proposed by Aristotle became a guideline followed by philosophers of the Lvov-Warsaw School. For Aristotle, all acts and decisions were always aimed at some good, as good was the goal of all endeavours.

Twardowski and his students, whose discussion of good and happiness was based on an analytical clarity of thought and language, continued, in a way, the Aristotelian moderation in proclaiming truths about good and happiness. They believed that the properties of acts and objects designated by the words 'good' and 'bad' were vested in objects about which such judgements were made in true statements. And judgements were statements which did not concern the objects to which they applied only apparently. This attitude, referred to as ethical objectivism, presumes the independence of good from human feelings, or the independence of the logical evaluation of a particular object from what is experienced by the judging subject. Such cognitive attitude in ethics is called intuitionism; its advocates also add that good is non-definable, and that it is an absolute value. And yet, intuitionists believe, absolute good is found in particular objects related in a multitude of ways to the knowing subject and to other objects. Therefore, absolute good should neither be identified with absolute, nor with the source or essence of existence. It should be recalled that intuitionism in ethics may take a number of forms. In the case of philosophers of the Lvov-Warsaw School, who were intuitionists in their understanding of moral good, we can distinguish between empirical and logical intuitionism, represented in the views of Tadeusz Czeżowski and presuming the existence of a certain axiological empiricism; empirical and emotionalist intuitionism represented by Tadeusz Kotarbiński, who believed that feeling was the essence of experience through which good was discovered; and, finally, intuitionism referred to as rationalist apriorism, present in the views of Tatarkiewicz. The precursors of that last form of intuitionism were Socrates and Aristotle, for whom the object of intuition were principles, primary systemic assumptions, general abstract truths of ethics.

The intuitive approach to the understanding of good in the Lvov-Warsaw School eliminated relativism in ethics. Relativism was criticised both by Twardowski as his students, who claimed that different behaviours displayed by people in different times and circumstances did not result from different judgements of what is right or wrong, but from different mores. Thus, a man following the custom does not necessarily do so because he believes that such behaviour is right or wrong. Due to cultural differences, identically worded ethical rules and judgements may have different connotations in various cultures; on the other hand, a rule or judgement which has the same meaning may be worded in an entirely different way. We should also note that according to philosophers of the Lvov-Warsaw School, ethical relativism must not be identified with subjectivism. Tatarkiewicz asserted, for example, that the relationship between relativism and subjectivism in the perception of good is most often caused by a similar way of thinking about good, and is therefore primarily a psychological, and not a logical, relationship.

The category of good, which fundamentally determines the understanding of happiness, was systemically analysed by two of the School's philosophers, Tatarkiewicz who, even though he was not one of Twardowski's students, was considered a representative of the School, and Czeżowski.

Good in the concept proposed by Tatarkiewicz is absolute, but not in the sense that it stands above all existence. Absolute good is good in itself, and not for someone. There are many goods because there are many things that have the property of being good, which is absolute and objective. These goods, according to Tatarkiewicz, include nobility, faith, pride, beauty, harmony, the fullness of life, happiness, joy, health (cf. Tatarkiewicz 1949).

A very interesting and inspiring view of the quality of being good or bad which Tatarkiewicz discusses in his paper 'O bezwzględności dobra' ('On the Absolute Nature of Good') is his assertion that an object which has some quality of goodness may not be good itself, and that an object which has some quality of badness may not be bad itself (Tatarkiewicz 1949: 65).

Of course, such an assertion may be used to justify a superficial charge against absolutism in ethics, as doubts arise about the fact that only some of the objects which possess the quality of goodness are good, while others are not. Tatarkiewicz defends his theory about the absolute nature of good by arguing that the unchangeable value of qualities may not be used to conclude that the value of objects themselves is unchangeable as well, or vice versa—the fact that the qualities of objects may change does not necessarily imply that qualities are changeable as well. Citing Max Scheler (1913), we might say that treacherous behaviour is always perceived as wrong, even if the person who has this quality gets rid of it, and friendship will always be good, even if our best friend betrays us.

Tatarkiewicz believed that the main task of scientific ethics, i.e. the kind of axiological views developed in the Lvov-Warsaw School, in particular by Twardowski, is to order, classify and arrange absolute values into a hierarchy; or, in other words, to determine the order of goods. Attempts at ordering values proposed by Tatarkiewicz do not form a fully-fledged system, however.

During an international conference held in Amersfootin in 1938, he presented a project of axiology which includes an ordering of values (Tatarkiewicz 1939). He divided them into five categories: (a) moral values (honesty, justice, kindness, nobility); (b) cognitive, or intellectual, values (truth, creativity); (c) aesthetic values (beauty and its varieties); (d) hedonic, or emotional, values (pleasure and its varieties); and (e) vital values, or natural goods (life, health, strength, good looks). This catalogue of values was supplemented with a subdivision into three basic classes, i.e. values proper to man, to objects, and those on the border of these two. His discussion of values is not a detailed study, however, and the conclusion which can be drawn from his rather curt comments boils down to saying that it is better to compare particular goods by applying specific principles of their classification, and that practically all types of values may coexist in a harmonious way. In his opinion, however, moral values stand at the very top of the hierarchy, and therefore should not be sacrificed for any other values. The structure of values in his views Tatarkiewicz corresponds to the structure of actually and potentially existing goods, since values do not exist without particular goods, even though they are absolute and objective.

The category of good in the views of Czeżowski is presented as the manner in which objects exist. Consequently, man-made evaluations lack any content, as in his opinion adjectives like 'good' or 'beautiful' add nothing to the description of an object. Judgements are, in fact, beliefs which are considered in terms of true or false. Thus, Czeżowski believes that good cannot be presented, just like there can be no presentation of existence; it can only be asserted (Czeżowski 1936). We can, however, present certain criteria of good, such as pleasure or moderation. Such criteria of good are not the same as good, however, since good is neither pleasure nor moderation. Goods, or values, may be absolute or relative. Goods are the object of man's endeavours, and we know that different people choose to pursue different goods (Czeżowski 1965: 130).

Thus, by reconstructing Czeżowski's axiological views, we may notice that in his opinion there exist so-called primary individual judgements, which may take the form, for example, of stating: 'This is valuable', and are often replaced by judgments of goods which have been doubly relativised: 'Something is good in view of something for someone'. He thus introduces into the logic he is constructing the notion of a 'valuating subject' and the notion of good in view of which something is valuable, which he refers to as the value parameter. The relationship between the object being judged as good and the value parameter—that is, the good and the valuating subject—is called a value-generating relationship.

Based on these axiological assumptions, Czeżowski provides a definition of absolute good: if the value parameter becomes identical with good, then good is valuable in the primary meaning of the word, and not in view of something else, but only in view of itself, then such good may be called absolute. The diversity of valuating subjects generates the plurality of goods, and consequently the plurality of value parameters and value-generating relationships.

This brings us to another problem, namely the relationship between individual goods and the methods employed to compare and arrange them. In Czeżowski's axiology, there are three value-generating relationships between goods: equality, superiority and inferiority, and the displacement of lesser goods by greater ones.

These relationships do not apply to absolute goods, however, since absolute goods are often considered to be incomparable. Even if considered as absolute, and thus incomparable, goods such as health, wisdom, or perseverance can be compared if instead of considering value-generating relationships as a whole, we only consider them in part. Which is why, according to Czeżowski, depending on which relationships are taken into account and which are left aside, either one or the other of the pair of goods which was initially incomparable will be considered greater, or both will be judged as equally valuable (Czeżowski 1965: 135).

Besides, taking into account comparable goods, we can identify the greatest and the least good, where the greatest will most likely prove to be that good which satisfies most of the valuating subject's needs, and the least will be that which only satisfies the minimum of his needs, in which case it can be replaced by any other good (Czeżowski 1965: 138). Apart from being compared, goods may also be added up. Czeżowski claims that the sum of two goods, where one is greater than the other, is equal to the one that is greater. He also presents the idea of creating a series of goods referring to certain parameters, and says that such parametrisation of goods may reveal the one we will find to be absolute. Naturally, all goods included in the series should be subordinated to the absolute good among them, identified in accordance with the logic of goods, and considered either as the means used to achieve this absolute good, or as its components.

We may thus say that absolute good is a parameter common to all goods. Czeżowski believes that absolute goods are objects which satisfy the criteria of values established by generalising individual empirical judgements (Czeżowski 1965: 138).

In most ethical concepts only one good is distinguished and considered to be absolute, even though theoretically there may be multiple absolute goods. However, eudaemonist-utilitarian systems, for example, consider happiness, understood as well-being or welfare, to be the only good; while perfectionist systems emphasise first of all perfection, which may be understood and interpreted in various ways. If, then, as is done in eudaemonism, or perfectionism, we reduce all goods to a single parameter, we will be able to compare goods which were initially difficult to compare. If we also perform a generalisation by eliminating so-called individual valuating subjects and replacing them with a 'subject in general', the only factor differentiating the relationships between the valuating subject and the objects which are called goods will prove to be value-generating relationships.

Czeżowski calls these relationships, firstly, relationships of entailment, and secondly, relationships of mutual independence, applying to non-comparable goods. Even here it is possible to compare the incomparable, however, by building an appropriate scale of goods, starting from the least of goods and ending with the greatest good, with the space between them filled with intermediate goods. So it is enough to properly grade the value-generating relationship between two goods which are considered to be incomparable. The place a particular good has on the scale of goods may thus, according to Czeżowski, show us quite objectively which good is greater and which is lesser in view of a given parameter. Therefore, goods which are commonly held to be incomparable can be compared, as we can identify the so-called common moment, which, being gradable, creates a scale of goods. With respect to utilitarian goods in particular, such scale is symbolised by money, which some, however, consider to be the measure of all goods.

To sum up the discussion of the category of good in the views of Lvov-Warsaw School philosophers, it should be observed they all agreed that good is an absolute value, and is therefore objective. There were, however, rather significant differences between the views held by Tatarkiewicz and Czeżowski. While Tatarkiewicz believed that good is a property of objects, Czeżowski asserted that good is a way in which objects exist. As regards the knowledge of good, Tatarkiewicz was of the opinion that it comes through intuition, understood as an autonomous type of cognition, while Czeżowski called intuition a kind of experience which may be improved.

Even though other philosophers of Twardowski's School, including Władysław Witwicki and Kotarbiński, discussed the problem of good in their reflections on ethics, they did not develop any fully-fledged theories on the subject. Witwicki proposed a thesis about the unity of truth and good (Witwicki 1936). He believed that just as there is one truth about the world which we arrive at when our cognition is based on experience and precise analysis, so there is one good—provided it is properly understood and defined (Witwicki 1936: 47).

For Witwicki, the primary, indeed, the only criterion of ethical judgement based on which ethical principles can be developed, is the voice of one's heart. It is the voice of one's heart that can detect any error or misunderstanding in judgement; it is on this voice that the development of morality depends. We can thus see that, compared to the precise axiological analyses performed by Tatarkiewicz or Czeżowski, Witwicki's views on the subject of good are quite superficial. Neither did Kotarbiński place any particular emphasis on analysing the category of good. Some believe that this is related to the reism he advocated.

It could be argued, however, that he understood good in three fundamental aspects. He did so based on three main tasks which he assigned to ethics. In his opinion, the first type of good provides for our happiness, and the related scientific reflection may be called felicitology. The second type of good helps us to effectively achieve our goals. This is praxeology, or the study of effective action. Finally, the third type of good is all that is respectable, equitable. This, Kotarbiński believes, is the subject matter of analyses performed by ethics proper (Kotarbiński 1979: 208–209).

3 The Category of Act

It is worthwhile noting that there were hardly any major differences among ethicists of the Lvov-Warsaw School in the way they defined act, which, being one of the basic axiological categories, was defined as the deliberate action of one who pursues a particular goal, such as happiness or perfection.

Significant controversies emerged in the School, however, when it came to analysing the moral evaluation of acts. The notion of act in ethics considered from the point of view of its evaluation represented a crucial and complex set of issues for this philosophical formation.

Tatarkiewicz explained and classified the notion of act on the axiological plane. Twardowski, Witwicki and Karol Frenkel analysed the problem of the moral judgement of act, while Marian Borowski and Kotarbiński considered it from the praxeological point of view.

In his reflections on act, Tatarkiewicz said that the absolute and objective nature of the theory of values entails, however, a relative and objective nature of the rules of just conduct. Why so? Because if an object having certain qualities is good, then any object having the same qualities is good. Yet if we consider the rules of just conduct and find that a particular act in certain circumstances is just, other acts having the same qualities may be, but are not necessarily, just (Tatarkiewicz 1949: 140). What does the justness of an act depend on, then? According to Tatarkiewicz, first of all on the acting subject, on the time, place, other people and things. We thus conclude that while an act is objectively just, the individual aspect of its justness needs to be emphasised, since an act is just only for particular subjects in particular circumstances. If we want to create a rule of just act, or a prescript which leads to obtaining the best desirable state of affairs, we must remember that it will only apply conditionally and on a case by case basis.

Consequently, such prescript must be constructed separately for each situation. Thus, every rule of just act must contain an empirical element, and be based on an inductive juxtaposition of individual instances of just acts or individual rules. Tatarkiewicz, like philosophers of the English school of analysis, claimed that only goods may be absolute, while rules are always relative.

Analysing the notion of act, Twardowski asserted that moral judgement does not apply to the act as such, but to the character of the man who performed it. The complicated network of circumstances, environments, living conditions affect decisions taken by the subject. Moreover, man's conduct becomes even more unique in view of his responsibility for what he does. Twardowski says that the responsibility of a moral subject cannot be avoided when: (a) that which he is being 'charged with' is the result of an act he performed; (b) the act was the result of a decision; and (c) the decision was consistent with the man's character (Twardowski 1971: 218). Twardowski foresaw that the development of ethics was going in the direction of judging actions and their consequences by judging the intentions behind them, and thus of judging man's actions in view of his character, or the permanent direction of his will.

Witwicki asserted that ethical evaluation applies first of all to decisions, or acts of will, which man's actions are the result of. He believed that one's will, design, decision are a necessary condition for an act to occur, and thus only decisions which are the primary 'manifestations' of will may be subject to judgement. When judging an act, one should first of all analyse the circumstances and motives for its performance. It is necessary to identify and examine all factors which may have contributed to the decision to act in a particular way. For we may often witness so-called involuntary acts which are not the result of inner motivation; in such case, their judgement will be different than in the case of acts performed by an effort of the will. The difference between such acts can be known, according to Witwicki, through introspection into man's inner states and character. It is this method that allows us to understand and evaluate the dynamics of human acts (Witwicki 1949: 41).

Twardowski and his oldest student, Witwicki, held similar views on determinism with regard to moral choices and the resulting actions. Twardowski believed that a sceptical argumentation concerning the freedom of action, suggesting that the moral subject does not have to account for his deeds because they are necessary and unavoidable, does not prevent moral conduct. For there is in man's consciousness a striving, Twardowski claimed, to a sense of free will, which, even though not necessary in substantiating the functioning of moral feelings, is required to generate them. Witwicki, on the other hand, believed that determinism does not remove responsibility for ones' deeds. He was of the opinion that punishment and reward significantly affect human behaviour. Besides, it may be conjectured that it is due to determinism that moral responsibility and the resulting praise or blame gain their proper sense. We may thus say that both philosophers allowed for the coexistence of determinism with a certain autonomy of human actions.

It is also worthwhile to look here at the views of Frenkel, one of the less known representatives of the Lvov-Warsaw School, who, like Twardowski and Witwicki, was convinced that it is decisions and not actions that should be the subject matter of ethical evaluation. In his papers 'O pojęciumoralności' ('On the Notion of Morality'; Frenkel 1925) and 'O przedmiotach oceny etycznej' ('On the Subject Matter of Ethical Judgment'; Frenkel 1919) he presented his views on the subject. He believed that it only appears to us that our actions and deeds are the subject matter of ethical judgement. In fact, it is the motives behind our moral or immoral decisions that are judged. Frenkel insistently defended determinism in his views as an undeniable principle governing reality, while leaving out the clearly basic ethical category of responsibility. Twardowski and Witwicki, on the other hand, tried to reconcile determinism with the fact that responsibility is that single value in ethics which cannot be left out. Therefore, contrary to Frenkel, they claimed that only after we become aware of the rules of determinism can we sensibly talk about responsibility, punishment and reward. They believed necessity does not cancel the ability to choose or the responsibility for choosing between right and wrong.

Axiological reflection on man's acts and deeds did not fully exhaust these immensely important problems studied by ethicists of the Lvov-Warsaw School. An interesting supplement was provided by their analysis of act in terms of its effectiveness. In this respect, we should discuss the views held by Kotarbiński. He called the problem of man's intentional action and its effectiveness praxeology, which he thought should be divided into three areas. The first one is the analysis of notions related to any deliberate behaviour. The second is based on a typology of actions, and concerns first of all criticism of the actual, practical ways of performing an action in view of their efficacy, potency, advisability, practicality. The third area is the so-called normative and advisory part, and includes guidelines for improving 'technical performance' (cf. Kotarbiński 1938: 612–613).

Kotarbiński's very pragmatic approach to the sphere of human actions was innovative for his times, and anticipated the contemporary reflection on the management of human resources, as with praxeology it is possible to identify and systematise various forms of action. 'Effective action is aimed at yielding a successful product immediately, without multiple failed attempts' (Kotarbiński 1938: 615).

In order for an action to be efficient, it must always be preceded with 'thought experiments', which helps avoid deformations and pathologies in the method employed. Providing a very broad definition of human action, Kotarbiński implied that man's actions can be considered in many aspects. Thus, actions include both thought and speech as well as deeds, conduct, active behaviour. For an action to occur, a system of inner and outer conditions enabling it must be in place. External feasibility can be achieved if no one and nothing other than the acting subject himself can prevent him from accomplishing what he intends. Inner feasibility exists if the acting subject has the necessary predispositions to perform the act. Kotarbiński believes these are: will, power, knowledge and ability.

At this point, Kotarbiński introduces the notion of 'performance', which is gradable, because the doers of deeds may perform better or worse in whatever they do. This quality depends on the extent to which the acting person was inventive, on how much effort and determination they put into the action, how creative they proved to be—as we would say today.

Kotarbiński called the energy, effectiveness and economy of an act its practicality, while the effectiveness and economy themselves its technical properties (Kotarbiński 1958a: 423). When analysing human actions, it is necessary to consider their properties which are aimed at achieving the intended goals. According to Kotarbiński, one must first develop inner 'resolve', which consists in the ability to constantly direct one's efforts at accomplishing a particular goal, despite obstacles making such constancy in action impracticable. Another important feature is the 'effectiveness' of measures taken to achieve the goal, pre-planned with strict deliberation. Yet another property is 'expertise', i.e. efficient, effective, appropriate action. According to Kotarbiński, the feature of 'expertise' may be evaluated on a scale from 'incompetence' to 'excellence'. Finally, still another feature is 'economy', consisting in a skilful use of resources through efficient and economical action.

Some very interesting reflections on strategic action can be found in Kotarbiński's essay, published in Warsaw in 1938, under the meaningful title 'Z zagadnień ogólnej teorii walki' ('Issues in the General Theory of Struggle'; Kotarbiński 1958b). Even before World War II he had come up with an idea, still inspiring management gurus today, which pointed to the analogy between efficient, competent action and struggle, understood of course not in the military sense but as an intentional striving of people acting, pursuing a particular goal out of different motives and using different means. Kotarbiński lists as many as twelve rules, such as the rule of anticipation, also called the rule of accomplished facts, or the 'tip of the scale' rule (Kotarbiński 1958b: 597), which consists in taking advantage of the economic tailspin of entities participating in the action who mutually neutralise their voices or forms of action. This rule allows the smartest to win. Yet another, equally inspiring rule still applied today, is the directive of gathering intelligence about the adversary and deliberately misleading the other party (Kotarbiński 1958b: 602).

Kotarbiński believed that a rational, reasonable adversary follows predictable patterns, while it is very difficult to predict the actions of a fantast who is not restrained by rational limitations. Thus, the greater the mastery in action, the less there is room for improvising. Kotarbiński argues that it is therefore necessary sometimes to imitate the irrationality of our adversary and take actions which may appear foolish, as in result they may bring a positive effect, victory. Kotarbiński was an ethicist and praxeologist in one, which made him realise that praxeological problems are not the same as ethical requirements. Praxeology does not include either emotional evaluations or moral judgements, though certain praxeological statements and recommendations may raise moral concerns. The conditions of action which are by all means expedient are often directly contrary to reactions dictated by noble feelings.

4 The Category of Happiness

Once we know what good is, and how to act to do good, the next natural step is to talk about happiness as the goal to be achieved by doing good. Here, of course, we must reflect on whether cognitive judgements about the quality of our life and our sense of happiness depend on the passing days and months, or on the amount of positive and negative emotional experiences, or perhaps on a general sense of happiness. What does our happiness depend on, then? Is it our inner experiences and system of values, or external, objective conditions which we have hardly any control over, such as our place of birth, our sex, living conditions, surroundings, political system, historical events? How is a happy life created? Most ambiguous and divergent descriptions and prescriptions for happiness were offered already in antiquity, and are still developed in modern times. Which shows us that the problem of happiness is important enough to be considered timeless, while being difficult to precisely define and identify.

What advice can we find in the concept of ethics proposed by Tatarkiewicz and his student philosophers of the Lvov-Warsaw School? We should note that the views they held, even though based on certain apriorist and intuitionist assumptions which could be summarised in the assertion that good is an absolute value, also emphasise the thesis about the empirical foundations of ethics.

Consequently, ethics as an empirical science is based both on objective and on relative, individual rules of moral conduct. It should be noted that there is in their axiological views a certain realism combined with universalism, for they do not advocate any particular, privileged way of judging ethical values. Which is why ethical codes give way to a realistic approach to reality, and specific circumstances and situations significantly affect actual norms of conduct. It should also be emphasised, particularly in the views of Kotarbiński and his concept of independent ethics, that metaphysical, worldview or scientific substantiations do not determine ethical choices leading to just acts.

Due to the fact that philosophers of the Lvov-Warsaw School were not proponents of so-called moralising, they never created a complete picture of the personal model of a happy human being. Nevertheless, by collating their views we may try to reconstruct such model, based on their suggestions and hints on how a person should act so that his life and the life with him can be morally approved and become happy.

What moves to the foreground is the postulate of active and rational resistance to evil, since passivity and thoughtlessness are not only conducive to, but actually cooperate with evil. At this point it should be noted that Twardowski as the School's founder provided an example himself of how to actively counteract evil. This attitude was visible first of all in his work as a teacher. Twardowski emphasised a model of philosophical education which included both intellectual and moral development, bespeaking not only of a strength of thought, but also a strength of character, like that of ancient masters. 'Twardowski educated through teaching, developed characters by instilling moral principles in his students – faith in the existence of the absolute values of truth and good' (Sośnicki 1959).

Ethicists of the Lvov-Warsaw School referred both directly and indirectly to certain role models from antique culture. One of the ways in which they emphasised these views was by translating the works of ancient philosophers into Polish. Moral models in the views of Plato or Aristotle were not archaic for them, since the virtues of fortitude, love of truth, commitment to perfection in one's actions appeared as valid as ever, if not downright indispensable in the Polish state which was experiencing its rebirth as independence was regained in 1918 after

120 years of slavery. Next to the principle of active participation in the life of the resurgent Polish society in order to reduce evil, another equally important principle was universal kindness. Its author was Kotarbiński, but he derived some of his inspiration from the words of his teacher, Twardowski, who defined this principle as identification of one's own interests with the interests of others during one of his lectures on ethics (Twardowski 1907).

Consequently, refraining from hurting others, fulfilling one's commitments, showing gratitude for the good experienced from others, earnestness, dutifulness, punctuality, and even respect for one's adversaries should all be traits of the character of a kind man. Based on these guidelines, Kotarbiński promoted the model of a reliable guardian, or someone who puts the duty to help others before his own interests. His reliability, courage and altruism are, according to Kotarbiński, only the minimum characterising a decent man. And yet, even this minimum is very difficult to achieve, requiring a lot of effort and sacrifice.

Kotarbiński's model of reliable guardian, or someone who implements the precept of universal kindness, was considered an attempt at a secular depiction of the Christian principle of the love of one's neighbours. It emphasises the assumptions of humanism, tolerance, respect for the dignity of man and recognition of the human right to optimum conditions of development. Nevertheless, the universalism of moral precepts endorsed in this concept limited their applicability due to the lack of any reference to actual economic, political and social circumstances. His abstracting from reality raised a lot of concerns, including questions about whether one can be a good guardian in a bad cause, or whether one can be called a good guardian if he acts for selfish reasons. The moral judgement of a guardian thus depends not only on who takes care of whom and in what matter, but also on the social role they are performing.

Critical comments about the concept of a reliable guardian did not obscure the authentic values it emphasised, however. The advocates of this concept, on the other hand, believed that its practical implementation required heroism. The ideal of man proposed by Kotarbiński, as well as other philosophers of the Lvov-Warsaw School: Twardowski, Witwicki, Tatarkiewicz, or Czeżowski, presumed a heroic attitude resulting from a specific approach to values in the human world. They were convinced that the necessary condition for the existence of values is the personal effort of a man who, while not possessing these values, nevertheless keeps striving towards them. So possessing values means striving towards them, believing that they are a very real and important element of the human world. The personal effort of every individual is necessary to consolidate and promote values, for even in a most developed society man is never excused from working to acquire knowledge, create art, develop technology, organise and improve social life. Persevering in this effort often requires an attitude which may be called heroic. The need for such an attitude becomes even clearer when we realise that everything man has achieved and won for all his sacrifice and labour can be lost. Values in the human world are always endangered by people who are insensitive and negate the effort put into achieving them, cherishing their own comfort or cynically taking advantage of others. The greatest enemy of values are all acts of aggression, especially war.

Heroism emphasised in the concept of man reconstructed from the views embraced by philosophers of the Lvov-Warsaw School did not only result from their particular attitude to values, but was also due to the specific conditions in which it was to be propagated and implemented.

In the years 1918–1939, Poland was rebuilding its identity after 120 years of slavery. This difficult independence was given to Poles in a country devastated by the occupants, tormented by various political groups fighting for power; a country which lacked schools and where industry had to be rebuilt. It took the effort of the entire society for Poland to begin a normal existence once again after 120 of political, economic and social nonexistence. In this multi-national country, torn by constant contradictions, struggling to rebuild its industry, agriculture and education—life went on anyway.

Naturally, there was an urgent need for emphasis on the model of a Polish patriot, a Polish educator of the new generation, who could show others how to work and live in a free country. While everyone knew that during the years of captivity Poles never lacked patriotism, it did not always take the form which was actually needed in the circumstances. In the times of peace and newly regained freedom, it was no longer sacrifice of one's life that was needed, but wise action and honest work.

In his essay 'O patriotyzmie' ('On Patriotism') Twardowski (1919) wrote that patriotism does not only consist in the love of one's language, tradition, heritage and nation, but also the love of one's neighbour, of one's fellow countrymen. He believed that anyone who spoke and felt the Polish way in the Polish homeland—was a Pole (Twardowski 1919: 14–16). Solidarity must not only be embraced when the nation is faced with an enemy, but also during reconstruction of its devastated country. He postulated that for the sake of the new Poland, social relationships should be purged of egotism, particularism, and ill-conceived partisanship (Twardowski 1919: 17). He believed that schools, particularly universities, now had special tasks to accomplish, as they developed the highest intellectual values known to man. Twardowski thus wrote of the dignity of universities, which must not be restrained by intervention from either the state or the church (Twardowski 1933).

Czeżowski claimed that schools and the process of educating the society could only achieve the desired cognitive goals if science was not subordinated to institutions which placed restraints on its pursuit of truth (cf. Czeżowski 1933: 11). He asserted that a school which forsakes its scientific character becomes a bad school not only from the academic, but also from the state's point of view. He compared such a school to a rudderless boat, tossed about by waves of political interests and compromises, which every new party coming to power could treat as an instrument in educating young people in the spirit of its own ideology (Czeżowski 1933: 12).

The relationship between the teacher and the student cannot be based on falsehoods and understatements, as the resulting evil undermines faith in the authority of the teacher, i.e. of the person who is supposed to teach others how to achieve such values as truth and good. A society without teachers who understand and follow moral principles degenerates, succumbs to destruction, losing its ability to develop and improve. Twardowski, Witwicki, Kotarbiński, Czeżowski and Tatarkiewicz, teachers and educators of the young generation themselves, were aware of the responsibilities and duties of teachers.

The fact that, aside from reflecting on good as the fundamental ethical category, the Lvov-Warsaw School also analysed the notion of happiness, was due mainly to Tatarkiewicz, even if some views on the subject were also expressed by other representatives of the School, mentioned above. It was Tatarkiewicz, however, who wrote the famous essay 'O szczęściu' ('On Happiness'; Tatarkiewicz 1965) in which he included everything that can be objectively said about happiness. That is why happiness and ways of achieving it are presented in his essay from two different perspectives, two different points of view. The author of the so-called 'Encyclopaedia of Happiness' included both theoretical, historical, and definitional analyses and their practical implications. He did not only consider the problem of happiness on the ethical and normative plane, but also on the psychological, sociological and semantic one. This comprehensive, holistic work was aimed at providing a nearly complete, definite picture of happiness—a notion which has been provided with many divergent definitions in the history of human thought. The expressions Tatarkiewicz uses to define happiness are approximate, i.e. model, ideal.

From the psychological point of view, it is interesting to note that Tatarkiewicz's reflections 'On Happiness', Czeżowski's essay 'On Happiness', and Witwicki's 'Pogadanki obyczajowe' ('Moral Causeries'), all discussing the problem of happiness, were written during the German occupation of Poland in the years 1939–1944. Some may be surprised that essays on happiness were written at a time when people suffered terrible distress during the war, afflicted by appalling tragedies. Tatarkiewicz explained this apparent, he believed, contradiction by saying that one who is surrounded by misery is bound to think more about happiness, because it is easier to suffer dire circumstances when one turns their mind to positive things.

5 Conclusion

Summing up reflections on happiness in the Lvov-Warsaw School, we may conclude that a happy life is one which has meaning and value, which is an ordered whole of actions taken in pursuit of harmonised goals, and which must be realised following a certain model, an ideal that is to be gradually achieved. The philosophers discussed here believed that happiness is a state which accompanies effective action, even though not all desired goals may always be achieved. One should not become pessimistic, however, as failure in one area may be compensated for by success in another. Positive thinking and self-confidence supported by a realistic assessment of one's possibilities is the basis of effective action.

The axiological project of the Lvov-Warsaw School thus includes an analysis of problems which are still valid and relevant in our lives today. Committed to the clarity of thought and language, the School initiated a discussion on modern man who should never forget dignity and honour, honesty and the need to pursue truth, as well as reasonable patriotism. Their postulate of restraining egotistic desires may seem overly idealistic today, in a world of rampant consumption. Yet the School's belief that knowledge and education are a guarantee of performance in every sphere of life, and that man's value is measured both by sound theoretical and practical knowledge and a strong character based on virtues and erudition, still provides foundations for the intellectual elites of a modern country in the twenty-first century.

References

Czeżowski, T. (1933). *O stosunku nauki do państwa*. Warszawa.

Czeżowski, T. (1936). O przedmiocie aksjologii. *Przegląd Filozoficzny, 39*, 465–466.

Czeżowski, T. (1949). Etyka jako nauka empiryczna. *Kwartalnik Filozoficzny, 18*.

Czeżowski, T. (1965). *Filozofia na rozdrożu: Analizy metodologiczne*. Warszawa: Państwowe Wydawnictwo Naukowe.

Czeżowski, T. (1969). *Odczyty filozoficzne*. Toruń: Państwowe Wydawnictwo Naukowe.

Czeżowski, T. (1976). O etyce niezależnej Kotarbińskiego. *Studia Filozoficzne, 3*.

Czeżowski, T. (1977). Transcendentalia – przyczynek do ontologii. *Ruch Filozoficzny, 35*(1–2).

Czeżowski, T. (1989). *Pisma z etyki i teorii wartości* (P. J. Smoczyński, Ed.). Wrocław: Ossolineum.

Frenkel, K. (1919). O przedmiotach oceny etycznej. *Przegląd Filozoficzny, 1.*
Frenkel, K. (1925). O pojęciu moralności. *Kwartalnik Filozoficzny, 3.*
Kotarbiński, T. (1937). Rozmowa o rozterce. *Wiedza i Życie, 6.*
Kotarbiński, T. (1938). O istocie i zadaniach metodologii ogólnej (prakseologii). *Przegląd Filozoficzny, 1.*
Kotarbiński, T. (1958a). Czyn. In *Wybór pism, 1.*
Kotarbiński, T. (1958b). Z zagadnień ogólnej teorii walki. In *Wybór pism, 1.*
Kotarbiński, T. (1961). *Elementy teorii poznania, logiki formalnej i metodologii nauk*. Warszawa: Zakład Narodowy im. Ossolińskich Polskiej Akademii Nauk.
Kotarbiński, T. (1979). Zasady etyki niezależnej. In *Studia z zakresu filozofii, etyki i nauk społecznych*. Wrocław: Ossolineum.
Kotarbiński, T. (1987). Obraz rozmyślań własnych. In *Pisma etyczne*. Wrocław: Ossolineum.
Moore, G. E. (1919/2005). *Principia Ethica.* New York: Barnes and Noble Books.
Scheler, M. (1913). *Jahrbuch für Philosophie und phanomenologische Forschung.* Berlin.
Sośnicki, K. (1959). Działalność pedagogiczna Kazimierza Twardowskiego. *Ruch Filozoficzny, 19*(1–2), 24–29.
Tatarkiewicz, W. (1919). *O bezwzględności dobra.* Warszawa: Gebethner i Wolff.
Tatarkiewicz, W. (1937). Ce que nous savons et ce que nous ignorons des valeurs. *Actualites Scientifiques et Industrelles, 539.*
Tatarkiewicz, W. (1938). Dobra, których nie trzeba wybierać. *Kultura i Wychowanie, 3.*
Tatarkiewicz, W. (1939). Von der Ordnung der Werte. *Actualites Scientifiques et Industrielles, 539.*
Tatarkiewicz, W. (1949). *O bezwzględności dobra.* Warszawa.
Tatarkiewicz, W. (1965). *O szczęściu.* Warszawa.
Tatarkiewicz, W. (1971). *Droga do filozofii i inne rozprawy filozoficzne.* Warszawa: Państwowe Wydawnictwo Naukowe.
Tatarkiewicz, W. (1989). Trzy etyki: studium z Arystotelesa. In P. Smoczyński (Ed.), *Dobro i oczywistość*. Lublin: Pisma etyczne.
Twardowski, K. (1895). Etyka wobec teorii ewolucji. *Przełom, 18.*
Twardowski, K. (1907). O zadaniach etyki naukowej. *Przegląd Filozoficzny, 10.*
Twardowski, K. (1919). *O patriotyzmie.* Lwów.
Twardowski, K. (1927). *Rozprawy i artykuły filozoficzne.* Lwów.

Twardowski, K. (1933). *O dostojeństwie Uniwersytetu*. Poznań.
Twardowski, K. (1934). *O tak zwanych prawdach względnych*. Lwów: Skład Głównyw Ksiegarniach S.A. Książnica-Atlas.
Twardowski, K. (1971). *Wykłady z etyki* (p. 9). Etyka: O sceptycyzmie etycznym.
Twardowski, K. (1973). *Wykłady z etyki* (p. 12). Etyka: O zadaniach etyki naukowej.
Twardowski, K. (1974). *Wykłady z etyki* (p. 13). Etyka: Główne kierunki etyki naukowej.
Twardowski, K. (1992). *Wybór pism psychologicznych i pedagogicznych* (R. Jadczak, Ed.). Warszawa: Wydawnictwo Szkolne i Pedagogiczne.
Twardowski, K. (1994). *Etyka*. Toruń: Wydawnictwo Adam Marszałek.
Witwicki, W. (1904). *Analiza psychologiczna objawów woli*. Lwów: Towarzystwo dla Popierania Nauki Polskiej.
Witwicki, W. (1906). *Analiza psychologiczna pragnień jako podstawa etyki*. XXI: Muzeum.
Witwicki, W. (1936). *Rozmowa o jedności prawdy i dobra*. Lwów.
Witwicki, W. (1949). *Komentarz do Dialogów Lukiana*. Wrocław.

6

Interpersonal and Intertextual Relations in the Lvov-Warsaw School

Anna Brożek

1 Introduction

Let us assume that the Lvov-Warsaw School (hereafter: LWS) is a sum of two subsets: (1) of the scholars and (2) of their output. By 'interpersonal relations' we mean the relations between people, namely between elements of subset (1). We use the term 'intertextual relations' to refer to elements of subset (2) and their content. Both of these concepts: relations between philosophers and relations between the contents of philosophical texts require an explication. We will begin with an outline of such explication, and then move to some illustrations taken from the LWS.

This paper is a part of the project 2015/18/E/HS1/00478 financed by the National Science Center (Poland).

A. Brożek (✉)
Faculty of Philosophy and Sociology,
University of Warsaw, Warsaw, Poland
e-mail: abrozek@uw.edu.pl

A. Drabarek et al. (eds.), *Interdisciplinary Investigations into the Lvov-Warsaw School*, History of Analytic Philosophy,
https://doi.org/10.1007/978-3-030-24486-6_6

2 Influence and Its Necessary Conditions

2.1 Contact Between Philosophers

One of the most important interpersonal as well as intertextual relationships between philosophers and between their works is most often described by the category of influence. This also applies to representatives of the LWS. We read, for instance:

> Twardowski was a talented teacher [...] and he exerted, through his teaching, a powerful INFLUENCE on generations of young Polish philosophers, such as Jan Łukasiewicz, Kazimierz Ajdukiewicz, Stanisław Leśniewski (who, in turn, taught Alfred Tarski) and Tadeusz Kotarbiński. This influence regarded first of all matters of method: Twardowski laid emphasis on «small philosophy», namely on the detailed, systematic analysis of specific problems – including problems from the history of philosophy – characterised by rigor and clarity, rather than on the edification of whole philosophical systems and comprehensive world-views. (Betti 2016; cf. van der Schaar 2015: 162; Smith 1989: 313)

However, the concept of influence remains unclear. What are the necessary and sufficient conditions of influence? Let us assume that *A* and *B* are philosophers. In order for the influence of *A* on *B* to occur, *A* and *B* need to have contact with each other. Not every contact results in influence, but a contact is at least a necessary condition of influence. Philosopher *B* may have contact with philosopher *A* in many ways.

Let us consider a situation in which philosopher *B* is a student of philosopher *A*. In such a situation, *B* not only reads *A*'s papers but also puts questions to *A*, presents his own results to *A*, is examined by *A*, prepares theses under *A*'s supervision, etc. Let us call this kind of contact between philosophers as an 'active contact of philosopher *B* with philosopher *A*'. The teacher–student relationship may be active to varying degrees. The most intensive example of an active relationship is that of apprenticeship with a (scholarly) mentor. The process begins with performing simple exercises assigned by the mentor, then progressing to more complicated tasks, to gradually becoming more and more independent.

Active contacts between philosophers should be distinguished from passive ones. Philosopher *B* may have a passive contact with philosopher *A* when reading his or her works, while *B* enters into no active relationship with *A*.

Let us now distinguish, within active contacts, oral contact from written contact on the one hand, and bilateral from unilateral contact on the other. Oral contact consists in oral conversation between philosophers (bilateral oral contact) or just listening to live or recorded lectures (unilateral oral contact). Bilateral written contact consists in exchanging written messages (like correspondence), and unilateral written contact is when messages are provided only one way.

Contact between philosophers is only a necessary condition for one philosopher exerting influence on another. Such contact may result in nothing: causing no action or change in the people who come into contact with each other. Influence of philosopher *A* on philosopher *B* occurs only if the contact of *A* with *B* results in some actions or convictions of *B*. For instance, as a result of contact with philosopher *A*, philosopher *B* starts to accept a certain thesis, takes a certain problem into account or applies a certain research method.

2.2 Types of Influence

The influence of *A* on *B* may take place on different levels, including that of personal life. Sometimes it is difficult to separate these two spheres: the substantive and the personal. For instance, contacts with one's mentor may motivate important decisions concerning one's professional life.

Let us add that *B* may be more or less aware of *A*'s influence. Sometimes *B* follows *A* consciously: repeats, develops or criticises elements of *A*'s philosophy, knowing exactly whose philosophy it is. However, *B* may also discuss *A*'s theses or problems not being aware of whose theses or problems they are. In particular, thesis *T* may be presented by *B* as *A*'s thesis or as *B*'s own thesis. In the latter case, *C* may not even be aware of the existence of *A*'s influence.

Influence seems to be further gradable. We will show further on that there are different spheres of influence, and that the attitude of influenced persons towards these spheres may also be different. Generally, the more the concepts/terminology, problems, methods and theses of *A* are accepted by *B*, the bigger *A*'s positive influence on *B* is. The more *A*'s concepts/terminology, problems, methods and theses are modified or rejected by *B*, the more *A*'s negative influence on *B* is.

Assume now that philosopher *A* was a teacher of philosopher *B*, and that philosopher *B* was a teacher of philosopher *C*. In such a situation, it is sometimes said that *C* is *A*'s 'philosophical grandchild'. We would prefer to say that *B* is directly influenced and *C* is indirectly influenced by *A* (cf. Kotarbiński 1952: 180).

Let us assume that philosopher *A* influenced philosopher *B* in such a way that *B* adopted (or accepted) thesis *T* (or a concept/term, a problem or a method) of *A*. Let us further assume that *B* influenced *C* in such a way that *C* adopted the thesis *T* (or a concept/term, a problem or a method) from *B*. In such a way, *A* influenced *C* indirectly, through philosopher *B*.

Not every successor of a given philosopher was influenced by that philosopher, but, of course, succession is conducive to influence. As a side remark, let us note that the existence of direct and indirect succession, as well as direct and indirect influences in a given group of philosophers, is a necessary condition of calling such a group 'a (philosophical) school'. One may formulate this condition as follows: if philosopher *B* belongs to the school of philosopher *A*, then *B* is a direct or indirect successor of *A* and *B* is under the direct or indirect influence of *A*.

The above partial description of the components of a philosophical school does not provide a sufficient condition for being its member. For instance, it does not answer the question of whether influences inside the school must be positive, or whether they may be negative as well. It seems that in some philosophical schools, many members modified and developed the initial ideas and findings of the founder.

2.3 Resonance: Its Spheres, Nature and Manifestations

In what follows, we will only discuss substantive influence, which we will call 'resonance' for short. Resonance may occur in two spheres: the sphere of teaching and the sphere of views. In the sphere of teaching by philosopher *A*, one may distinguish:

- 'the spirit' of *A*'s teaching;
- *A*'s didactic methods;
- the contents of *A*'s teaching.

In the sphere of scholarship itself, in particular with regard to philosopher *A*'s views, one should distinguish:

- the conceptual scheme and terminology *A* uses;
- the questions *A* poses—i.e. *A*'s problems;
- the ways *A* answers these questions (and substantiates them)—namely *A*'s methods;
- the answers *A* provides—i.e. *A*'s theses and systems of theses.

In short, *A*'s influence on other philosophers may concern at least concepts/terminology, problems, methods or theses.

The issue of resonance becomes even more complicated when we realise that it may consist in three kinds of attitude to influencing spheres: approval, non-approval or even disapproval of elements of a person's thought. Approval occurs when *B* accepts *A*'s concepts/terminology, problems, methods or theses. Non-approval occurs when *B* modifies *A*'s concepts/terminology, problems, methods or theses. Finally, disapproval occurs when *B* rejects *A*'s concepts/terminology, problems, methods or theses.

If *B*'s reception of *A*'s philosophy is favourable, then we say that *B* is an adherent, follower or epigone of *A*. If *B*'s reception of *A*'s philosophy is not favourable (and consists in modification or disapproval), then *B* is called 'a critic' of *A*.

2.4 Identification of Influence

Looking at the distinctions made in the previous paragraphs, one may easily notice that the concept of influence is quite complicated: there are many types of influence. The same is true for manifestations of influence, which brings us to the following problem: How can we know that one philosopher influenced another? The question is not easy to answer. In order to identify succession between philosophers, it is necessary to know relevant facts about their lives. Based on their biographies, we may determine whether there was any contact between them, and what kind of contact it was, if any (active or passive, oral or written, etc.).

In the case of passive relationships and types of influence, more resources are available. One may learn about *A*'s influence on *B* based on:

a. *B*'s declaration (i.e. *B*'s stating that *B* was influenced by *A*);
b. quotations of *A*'s works (or other references to *A*) in *B*'s works;
c. the occurrence of concepts/terminology/problems/methods/theses used by *A* (or their modifications, etc.) in *B*'s works.

Let us strongly emphasise that conditions (a), (b) and (c) are only conducive to influence: they are neither its necessary nor sufficient conditions.

Consider condition (a). Is a declaration by *B* sufficient to assume that influence occurred? It seems not. Firstly, philosopher *B* may be insincere in his or her declaration (for instance, *B* may declare to be under the influence of *A* because *A* is a fashionable or popular philosopher, or because it is required of *B* to be under the influence of *A*). Secondly, philosopher *B* may be mistaken in their claim of being influenced by *A*. In order to be influenced by *A* (in a positive or negative way), one must at least understand the content of *A*'s work. And it is often the case that influence is only apparent due to a lack of understanding.

On the other hand, the fact *B* does not declare him- or herself to be influenced by *A* does not prove no influence occurs. Firstly, *B* may not be aware of *A*'s influence. Secondly, *B* may not want to admit (including him- or herself) that *B* is under the influence of anyone.

Let us look at condition (b). Quotations may be, but are not necessarily a testimony of the existence of influence. Firstly, expressions in the form: 'see Smith 2015' often appear in a text not only because the author accepts Smith's thesis, but also when the quoting author has formulated similar theses or posed similar questions independently of Smith. We cite a philosopher not only in order to refer to his or her views, but also because we realise that this is required, for instance, by the reviewers of our paper. Moreover, a situation described by Chen Chung Chang in reference to Tarski may be the case:

> I will not make a point of mentioning Tarski's name each time his influence is either directly or indirectly present in this paper. This is because his influence in model theory is felt everywhere, and to mention him every time there is a connection with the subject discussed here would be an exceedingly monotonous exercise. (quoted after [Woleński 1989: 183])

Let us finally analyse condition (c). To an even greater degree, the presence of *A*'s theses (or their negations or modifications) alone in the works of *B* is not a sufficient condition of *A*'s influence on *B*. It may be the case that *B* reaches the same conclusions as *A* independently, not having had any contact with *A*.

3 Influences Within the LWS: Examples

The conceptual scheme sketched above is a proposal of how to understand the issue of influences between philosophers. In the following sections, as mentioned in the introduction, we will look at the phenomena discussed above as illustrated by the example of LWS. As the task of presenting the entire map of interpersonal and intertextual relationships in such a large group of scholars would exceed the scope of this paper, we will limit ourselves to some examples.

We will begin by describing the contact between the founder of the LWS and his students, as well as some of the philosophers and teachers of the school's first generation. Then, we will move to substantive influences in the sphere of ideas. Once again, we will limit ourselves to relationships between Twardowski and a few of his direct students, and only to some examples of various intertextual relationships.

3.1 Contacts Between Twardowski and His Students

The founder and the *spiritus movens* of the LWS was Kazimierz Twardowski. He is renowned as an excellent teacher and a kind of professor of philosophy, who saw his didactic tasks as his life's mission.

The impression Twardowski made on his students and everyone he met was often recalled. Stanisław Łempicki wrote:

> At 8 a.m. sharp, at the crack of winter dawn, throngs of students are crowding into the university building at St. Nicholas Street for Twardowski's lecture. The large hall is filled to the brim, dim gas light illuminates the notes [...]. The master, still young – he is not even 40 years old yet – [...] solemnly approaches the desk. The lecture is like a religious service; everything is listening attentively to these clear words, often about difficult issues; only the creaking of pencils can be heard. (Łempicki 1938: 31)

And here is Stanisław Lam's description of Twardowski as a lecturer:

> He was tall, broad-shouldered, with a raw face, surrounded by a long, light red beard. He wore a soft black hat, a black tie tied to a knot, a coat of military fashion, with horn buttons and a flap in the back. He had an upright bearing, walked slowly, majestically, always holding a leather-bound book under his arm. He never parted with it – even when he went from his house to his favourite Scottish Café. He gave lectures at seven o'clock in the summer. [...] He talked about logic, about the history of philosophy, about philosophical systems and currents of mind as easily and clearly as if he were talking about the most ordinary things. He was able to popularise the most intricate issues. (Lam 1968: 36)

The lecture hall was a place of unilateral oral contact between Twardowski and his listeners. His seminar was the centre of bilateral relationships in both oral and written form. In the following, often quoted, passage, Dąmbska described the perfect organisation of Twardowski's seminar:

> The seminar was a meeting place for students in all years of study, beginning with the second. It was there that under the direction and most careful attention of the professor they prepared for independent scholarly work. Classic philosophical works were read and interpreted together (always in the original, which required the knowledge of foreign languages). [...] Twardowski provided his students with ideal conditions for work. Every participant in the seminar could use the reading room, to which they had their own key: from 7 a.m. to 10 p.m. In that reading room they had their own desks and could avail themselves of the required books from a large library (which in 1930 had some 8000 items). [...] Everyone had the right and opportunity to contact the professor at will; he received students in his office, day in day out, between noon and 1 p.m. He used to spend some eight to nine hours daily at the seminar, often visiting the reading room and having many personal contacts with the participants. Such was the outer framework of his work, which was a unique educational activity of its kind. (quoted in [Woleński 1989: 4])

Twardowski's attitude towards participants in the seminar was described by Stefan Baley:

> Objections and counterarguments were presented by the professor [...] carefully, without any special pressure, so that we felt as though it was not him speaking to us then, but that it was truth itself speaking through his lips; truth which demands that everything be illuminated from all sides and that all possible arguments for and against be taken into account. (Baley 1936/1937: 66)

Besides the seminar meetings, Twardowski met students at the Philosophical Circle. The significance of this institution and of the role Twardowski played in it was described by Jan Łukasiewicz in his 'Diary':

> The meeting of the Philosophical Circle would start with a paper presented by one of the students, and then a discussion followed. Everybody waited for what Twardowski would say. One believed that he

> was able to resolve any problem. And there were many problems to solve. Does a man have a soul, does a man always act egoistically, can one say based on the style of writing whether a work was written by a man or a woman, etc. [...] The Philosophical Circle was an excellent school of thinking and had a great impact on youth. It was thanks to the Circle that I moved from law to philosophy and became a student of Twardowski. (Łukasiewicz 2013: 34)

3.2 Twardowski's Resonance in the Didactic Sphere

The spirit, style and methods of Twardowski's teaching were adopted by his students. Among the direct successors of Twardowski, there were relatively many significant teachers, impacted by his didactic spirit. Let us mention three talented teachers in the first generation of the Lvov-Warsaw School.

Let the first example be Kotarbiński. The circle of his Warsaw students may be compared to Twardowski's circle in Lvov. Kotarbiński spoke clearly and emphatically, infecting other people with his passion for philosophy. He published many papers of an educational character and influenced representatives of many disciplines. He wrote:

> I understood my mission [...] as preparing future teachers, and a teacher is any person who presents theses, independently of whether they do so it as a school educator or as a scientific researcher, as an author or a journalist. (Kotarbiński 1973: 144)

He paid special attention to philosophy and its weaknesses. Just like Twardowski, he was convinced that the main sin of philosophers is the obscurity of style. He tried to use his programme of semantic reism to bring clarity and simplicity to philosophical papers. Despite the fact that the programme proved to be impossible to implement, it impresses its stamp on everyone who has tried to follow Kotarbiński. The style of Kotarbiński is easily cognisable and his circle is sometimes called the 'school of *Elements*' (from the title of Kotarbiński's main textbook [1929]). Kotarbiński wrote about another great teacher, Ajdukiewicz, that 'He was a master of subtle and deep criticism, and thanks to his criticism as well as excellent advice he was very helpful to students' (Kotarbiński 1967: 81).

Ajdukiewicz taught philosophy in Lvov, Warsaw, Lvov again, Poznań and Warsaw again. In all these places, he proved to be an excellent teacher and organiser of philosophical life. One of his students and followers, Klemens Szaniawski, wrote:

> Kazimierz Ajdukiewicz was a professor by vocation. Not just a scholar, but a professor. He liked teaching, and didactics was one of the important stimuli of his work. While preparing lectures, he asked himself the questions which were his scholarly work. He used to say jokingly about himself that he had done nothing in his life but write a textbook on elementary logic. (Szaniawski 1983: 12)

Izydora Dąmbska was one of the youngest female students of Twardowski. She was unusually gifted as a pedagogue and had a great didactic passion, to the same degree as Kotarbiński and Ajdukiewicz. It was very unfortunate that she was twice denied the possibility to teach at university due to political reasons. Even in the short period when Dąmbska was a professor of philosophy at Jagiellonian University, collaborating with Roman Ingarden, she managed to create a school and educate a group of faithful students. One of them, Jerzy Perzanowski, wrote about her (paraphrasing the words of Dąmbska about Twardowski):

> Her great heart, full of deep emotion, from which we all drew, made Dąmbska a hunter of souls similar to Socrates. This heart created inseparable bonds between students and the master, and between one student and another – the bonds of friendship. And this is why Dąmbska may be called a master of our times. (Perzanowski 2009: 22)

3.3 Examples of Twardowski's Theoretical Resonance in His Disciples

Let us consider various forms of theoretical influence that Twardowski had on seven of his direct students, namely: Władysław Witwicki,

Łukasiewicz, Stanisław Leśniewski, Kotarbiński, Tadeusz Czeżowski, Ajdukiewicz and Dąmbska. There are some current opinions about Twardowski's theoretical influence on them, but many questions are still open: Did they openly admit Twardowski's impact? Did they cite his teachings? What was the sphere of Twardowki's influence (did his students take only problems and methods from his teachings or the answers as well)? What was the mode of influence: were they followers or critics of Twardowski? We will provide some partial answers to these questions based on theoretical papers by the above philosophers.[1] Then, we will also briefly discuss the influence in the student-to-teacher direction. Finally, we will discuss the question why there were such deep differences between various representatives of Twardowski's school. It seems that the special features of Twardowski's resonance may serve as an explanation of this fact.

3.3.1 Witwicki

The correspondence between Witwicki and Twardowski certifies the fact that the former was one of the closest successors of the latter in many ways: in his didactic programme, theoretical results, evaluation of other philosophers, including other representatives of the LWS and—last but not least—in the sphere of worldview. Twardowski confessed that he and his student 'had a heartfelt relationship' (Twardowski 1931: 454).

In his research, Witwicki faithfully followed the problems, conceptual scheme and terminology used by Twardowski. He also applied Twardowski's analytical method, which Czeżowski called 'analytic description' (Czeżowski 1953a). In some particular solutions of the analysed problems, Witwicki differed from Twardowski (for instance, in his analysis of decisions as judgements about future acts) and he proposed many

[1]We will omit texts about Twardowski's personality and papers written by Twardowski's students on special occasions (cf. Witwicki 1920, 1938; Kotarbiński 1936; Czeżowski 1948; Ajdukiewicz 1959).

original concepts, of which the best-known is probably the so-called kratism theory.

Witwicki did not hesitate to openly admit the existence of a theoretical connection with Twardowski. The traces of direct theoretical resonance may be found in many of Witwicki's papers, for instance, in (Witwicki 1900, 1904, 1905, 1925, 1931, 1932, 1939, 1947, 1957), as well as in his comments to translations of Plato's dialogues.

3.3.2 Łukasiewicz

The currently prevalent opinion is that Łukasiewicz departed from Twardowski in the theoretical sense: beginning his work in the domain of mathematical logic, he refuted all of his earlier works written in the spirit of the Lvov School. However, Łukasiewicz refers to Twardowski in his papers, and surprisingly often does justice to his teacher. In the paper 'O twórczości w nauce' ('On Creativity in Science'), Łukasiewicz remarked that it was Twardowski (1901) who used the term 'rozumowanie' ('reasoning') in order to refer to both 'wnioskowanie' ('inference') and 'dowodzenie' ('proof'), and that his own (Łukasiewicz's) views were a development of his teacher's approach (Łukasiewicz 1912: 16). He also admitted taking the term 'rules of reasoning' (Łukasiewicz 1912: 19) from Twardowski. A similar remark may be found in the paper 'O rodzajach rozumowania. Wstęp do teorii stosunków' ('On Types of Reasoning. An Introduction to the Theory of Relations') (Łukasiewicz 1911: 232).

In the paper 'O znaczeniu i potrzebach logiki matematycznej' ('On the Significance and Needs of Mathematical Logic'), Łukasiewicz assigns to Twardowski the role of the founder of the Warsaw School of Logic. He wrote:

> Almost all philosophers who practice mathematical logic are students of Professor Twardowski; for they belong to the so-called "Lvov School of Philosophy" where they were taught to think clearly, diligently and methodically. Thanks to that, Polish mathematical logic achieved a higher level of precision than mathematical logic abroad. (Łukasiewicz 1929: 426)

Łukasiewicz's declarations in his letters to Twardowski are even more interesting in this respect. Let us quote the three most characteristic paragraphs:

> February 6, 1905. I know very well that you do not share my standpoint [concerning the tasks of philosophy]. However, by working in a direction different from yours, I still feel very well that I am, even in this area, your student. This interest in scholastics and Aristotle that you evoked in me with your lectures on the history of ancient and medieval philosophy, [...], and first of all this scholastic element *par excellence* in reasoning and discussing which became my training in logic, all of that may have led me to the place where, today, [I am]. (Łukasiewicz 1998: 470)
>
> March 30, 1925. Step by step, I am moving away from philosophy. I am aware now that this started already in Lvov, not without your influence, dear Professor, as you taught us to think scientifically. (Łukasiewicz 1998: 487)
>
> July 14, 1931. Today, I feel rather like a mathematician than like a philosopher and I feel more connected to mathematics than to philosophy. I often recall what you, dear Professor, told me thirty years ago, after my final exams in mathematics: "Please do not neglect mathematics". (Łukasiewicz 1998: 494)

3.3.3 Leśniewski

Stanisław Leśniewski's attitude to Twardowski is illustrated by the dedication of his work 'O podstawach matematyki' ('On the Foundations of Mathematics'): 'To my esteemed and beloved Professor of Philosophy, Kazimierz Twardowski, Ph.D., I dedicate this work as a delayed tribute on his jubilee. A philosophical apostate, but a grateful student' (Leśniewski 1992: 174).

For Leśniewski, his apostasy from Twardowski's philosophy preceded his apostasy from philosophy as such. In the paper 'Przyczynek do analizy zdań egzystencjalnych' ('A Contribution to the Analysis of

Existential Propositions'; Leśniewski 1992: 8), Leśniewski refers to Twardowski's concept of connotation presented in 'Zur Lehre' (Twardowski 1894) in a rather neutral tone. Yet only three years later, in the paper 'Krytyka logicznej zasady wyłączonego środka' ('A Critique of the Logical Principle of the Excluded Middle'), he uses 'Professor Twardowski's "object of general representation of a triangle"' as an example of a philosophical myth (Leśniewski 1992: 50–52).

3.3.4 Kotarbiński

Ajdukiewicz called Kotarbiński and Czeżowski 'the most faithful followers of the programme of explaining concepts' formulated by Twardowski (Ajdukiewicz 1937: 252–253). In fact, both of them implemented this programme.

Tadeusz Kotarbiński acknowledged Twardowski's influence in the 'Introduction' to his textbook 'Elementy teorii poznania, logiki formalnej i metodologii nauk' ('Elements of the Theory of Cognition, Formal Logic and the Methodology of Sciences'). He wrote:

> [...] I wish to mention Professor Kazimierz Twardowski, my principal teacher during my university studies. I should be glad to win his approval for the opinion, expressed in this book, concerning the scope of epistemological knowledge required for further teachers. I should also be glad to come close to the model of simplicity, clarity and intelligibility of exposition which I was provided with by all of his works. I should also be glad to contribute to popularising his didactic, and not only didactic, principle, which recommends that we should be very particular about the meanings of the words we use, and thus avoid vagueness in both thought and speech. (Kotarbiński 1929: XII)

As we already mentioned, the programme of semantic reism presented in 'Elementy' was Kotarbiński's idea of how to give clarity to our ideas. An important difference between him and Twardowski lies in ontology: Kotarbiński accepted metaphysical reism (the thesis that only things exist), while Twardowski's ontology was rich.

Kotarbiński also continued Twardowski's idea of independent ethics. In his praxeology, Kotarbiński often used the distinctions introduced by Twardowski (for instance, in the paper 'Actions and Products' [Twardowski 1912]). However, Twardowski's influence was not always acknowledged by Kotarbiński in the form of citation. In 'Elementy', there are only a few references to his teacher. An example of explicit reference to Twardowski may be found in Kotarbiński's criticism of non-rational relativism. Kotarbiński wrote:

> By "relativism" one [sometimes] understands the view that scientific hypotheses should be changed, because a hypothesis which is compatible with all facts known at a given time turns out to be incompatible with them as soon as new facts, not known before, are discovered. There are also other forms of relativism which are difficult to argue with. We have to be careful, however, not to identify them with this irrational relativism against which [...], and in accordance with Twardowski, the edge of our criticism was directed. (Kotarbiński 1951: 300).

3.3.5 Czeżowski

In currently prevalent opinion, while Czeżowski was one of the closest students of Twardowski, the important difference between them was that Twardowski considered psychology as the methodological basis for philosophy, while for Czeżowski, this role was played by logic. In fact, in the 1920s, Czeżowski proposed an original modification of Twardowski's views by expressing them in the language of modern logic. It his paraphrase, he spoke of individual terms and variables instead of individual and general names: 'Just like a variable – a general presentation does not refer to a single, individual, strictly defined object, but to any object which has the properties indicated in the presentation' (Czeżowski 1925: 109).

He also argued that the concept of issued judgement and auxiliary presentation, which is an element of presented judgement, may be replaced by a sentence, a propositional function and an appropriate variable: '[The latter] is usually the product of simple propositional functions joined by the same variable [so] it is a complex of presented judgements referring to the object [...] of the auxiliary presentation' (Czeżowski 1925: 109).

Many years later, Czeżowski revisited Twardowski's text 'O jasnym i niejasnym stylu filozoficznym' ('On Clear and Unclear Philosophical Style'; Twardowski 1920), and challenged his teacher's standpoint. Twardowski's view was that if people speak unclearly, then they also think unclearly. Czeżowski drew attention to the fact that lack of clarity is sometimes a function of both the quality of the sender's thought and the intellectual competence of its receiver (Czeżowski 1954: 255ff.).

In the English edition of his selected papers—'Knowledge, Science and Values'—Czeżowski refers to Twardowski explicitly only once: mentioning his teacher's paper 'O istocie pojęć' ('On the Essence of Concepts'; 1903) as an example of the application of the method of analytic description. There are no other explicit references to Twardowski; however, a careful reader will easily notice that the papers 'On the Problem of Induction' (Czeżowski 1953b) and 'The Classification of Reasoning and Its Consequences in the Theory of Science' (Czeżowski 1963) develop Twardowski's results presented in the (unpublished) lectures 'Teoria badań indukcyjnych' ('The Theory of Inductive Studies') and in the textbook 'Zasadnicze pojęcia dydaktyki i logiki' ('Basic Concepts of Didactics and Logic') (Twardowski 1901). Moreover, the main theses of Twardowski's paper 'Pesymism i optymism' ('Pessimism and Optimism'; 1899) were expressed by Czeżowski in his text 'O szczęściu' ('On Happiness') (Czeżowski 1940). One may wonder whether these evident similarities are only coincidences or rather the result of long-lasting contacts between Twardowski and Czeżowski, and should therefore be analysed in terms of influence.

3.3.6 Ajdukiewicz

Also Ajdukiewicz, a close student of Twardowski, and privately his son-in-law, referred to his teacher on very rare occasions (except for the papers prepared to present Twardowski's personality or output). For instance, in the two-volume collection of papers 'Język i poznanie' ('Language and Cognition'), there is only one, purely theoretical mention of Twardowski. In the paper 'O pojęciu istnienia' ('On the Notion of Existence'), Ajdukiewicz refers to him as an example of a philosopher who accepts a mode of existence different from the real one (Ajdukiewicz 1951: 145).

3.3.7 Dąmbska

The situation is different in the case of Izydora Dąmbska, one of the youngest and closest students of Twardowski. For instance, in the English volume of Dąmbska's selected papers 'Knowledge, Language and Silence' (2015), there are several explicit references to Twardowski. Let us look at some examples.

In the paper 'Conventionalism and Relativism' (1934), she accepts Twardowski's method of refuting the arguments proposed by relativists (Dąmbska 2015: 110); the distinction between conventionalism and relativism; and the 'paradoxical' thesis that conventionalism may be useful in overcoming relativism (Dąmbska 2015: 112). She also cites Twardowski's comment about the ambiguity of the notion of relativity in some relativists (Dąmbska 2015: 114). In the paper 'On the Semantics of Conditional Sentences', prepared in honour of Twardowski, Dąmbska presents his views about conditional propositions (Dąmbska 2015: 263). She also mentions the fact that Twardowski drew her attention to the possibility of a twofold interpretation of the form: 'If we perform a surgery, the patient will not recover' (Dąmbska 2015: 270). In the paper 'When I Think of the Word "Freedom"' (Dąmbska 2015: 345), Dąmbska expresses the view that the relational character of the word 'freedom' was probably what Twardowski meant in his lecture 'Etyka i prawo karne wobec zagadnienia wolności woli' ('Ethics and Criminal Law in Consideration of Free Will'; Twardowski 1904/1905: 131). Finally, in the article 'On the Semantics of Adjectives', she notes that Twardowski rightly states in 'Zur Lehre' (Twardowski 1894):

> One of the sources of confusion between the notions of content and object of presentation, so common in logic and psychology, is the lack of

> distinction between the modifying and the determining function of the adjectives "presented" or "imagined". (Dąmbska 2015: 243–244)

However, Dąmbska criticises Twardowski's supplementation of Bretano's division into determining and modifying adjectives, proposed in his paper 'Z logiki przymiotników' ('On the Logic of Adjectives'; Twardowski 1927). Twardowski introduced two other kinds: reinstating adjectives and abolishing adjectives (Dąmbska 2015: 244). After providing a detailed presentation of Twardowski's modification, she comments:

> What place should the function of the adjective "supposed", as in the expression "a supposed object", be appointed in Professor Twardowski's classification? If the Professor agreed that this expression is not devoid of meaning, it is not my intention to resolve it at this point. What is, then, the purpose of distinguishing abolishing adjectives? It appears that even if we accepted that the expressions to which adjectives are attached are equivalent to the corresponding negative names, which seems wrong except for the expression "a supposed object", we would still have to assume that their function is not limited to the removal of the meaning of the name. [...] Therefore, it seems that, based on Professor Twardowski's classification, the group of [abolishing] adjectives should be included in the class of modifying adjectives, perhaps with the exception of the adjective "supposed" in the expression "a supposed object"'. (Dąmbska 2015: 246–247)

It is worth adding that in the footnote to the paper 'Prawa fizyki wobec postulatu prawdziwości twierdzeń naukowych' ('Laws of Physics and the Postulate of Truthfulness of Scientific Statements'), Dąmbska reveals that the text was amended based on comments by Twardowski (Dąmbska 2015: 158).

3.3.8 Twardowski's Assessment of His Influence

Finally, let us add that Twardowski was fully aware that his students had adopted many of his ideas, and did not care about being cited or acknowledged by them. He was simply interested in solving problems by making even small steps forward in difficult philosophical matters and by sharing these results with those around him. He preferred oral to written contact. He published very little and was happy that his thoughts were assimilated by his listeners.[2] In his autobiography, Twardowski wrote: 'Although in this way the results of my work do not always reach wider circles under my name, I do not regret it, especially since the printed word will not avoid a similar fate' (Twardowski 1926: 32).

4 Theoretical Resonance of Twardowski's Disciples in Himself

While theoretical resonance in the teacher-to-student direction is natural, resonance in the opposite direction, students influencing the teacher, is quite rare. It is interesting to note that such resonance occurs within the LWS.

Besides purely bibliographical notes (included, for example, in the paper 'O psychologii, jej przedmiocie, zadaniach, metodzie, stosunku do innych nauk i o jej rozwoju' ('On Psychology, Its Subject, Method, Relation to Other Sciences and Its Development'; Twardowski 1913), from among students listed in the previous paragraph, Twardowski refers in the collection of his selected works[3] to the following scholars: Witwicki, Łukasiewicz, Leśniewski and Kotarbiński. He indicates Witwicki as the one who 'places judgment and conviction completely on a par with pain and joy by speaking not only of experiencing pain and joy, but even of experiencing a conviction' (Twardowski 1912: 110); cf. Witwicki's 'W sprawie przedmiotu

[2]Not all of Twardowski's students followed him in this respect. For instance, Leśniewski was very sensitive to the problem of plagiarism. It was one of the reasons for his conflict with Ajdukiewicz.

[3]'Wybrane pisma filozoficzne' ('Selected Philosophical Papers'; Twardowski 1965); 'On Actions, Products and Other Topics in Philosophy' (Twardowski 1999).

i podziału psychologii' ('On the Subject Matter and Classification of Psychology'; Witwicki 1912). He makes some critical comments on Łukasiewicz's terminology; in particular, he criticises the fact that Łukasiewicz uses the term 'judgement' in reference to what Twardowski calls 'sentences'.

> Professor Łukasiewicz does so, among others, when defining judgment as a "sequence of words or other signs which state that some object does or does not have a particular attribute" (Łukasiewicz 1919: 12). But in treating judgment as a sequence of words or other signs, Łukasiewicz must distinguish from this sequence of words or other signs that which constitutes its meaning. As a matter of fact, Łukasiewicz also speaks of "meaning-equivalent judgments", defining them as judgments that "express the same thought in different words" (Łukasiewicz 1919: 15). Now, this thought, expressed in those words, is obviously nothing other than a judgment in the sense of a product of the action of judging; thus, if the word "judgment" is made to serve for designating "a sequence of words or other signs" that express this sort of thought, an expression will then be lacking for designating such a thought. (Twardowski 1913: 129–130)

However, Twardowski also notes that he took from Łukasiewicz (1904), the theory of relationships between judgements (Twardowski 1914).

In another paper, Twardowski presents a discussion between Leśniewski and Kotarbiński concerning the theory of truth:

> The objective value is what we mean by saying that events from the past cannot be changed. Does this apply to the future as well? Is it true that what is going to happen must happen as well? This problem was dealt with by Tadeusz Kotarbiński, Ph.D., in his dissertation. [...] [His thesis is as follows:] "Every truth is eternal but only some truths are both eternal and sempiternal". [...] Kotarbiński's idea was subjected to criticism. Among others, Stanisław Leśniewski, Ph.D., responded to it in his paper (Leśniewski, 1913). [...] I just wish to point out some DOUBTS concerning Kotarbiński's arguments". (Twardowski 1923/1924: 251–253)

Generally, Twardowski advocates cautiously on the Leśniewski's side.

5 Diversity of the LWS and Its Sources

Among the historians of philosophy, there is a belief that Twardowski's students were not related by any specific philosophical ideas, and especially not by worldviews. They included priests and atheists, conservatives and socialists, ontological monists and advocates of rich ontologies, axiological absolutists and axiological relativists, followers and opponents of many valued logic.

There are a few explanations of this fact.

The first is connected with Twardowski's didactic programme. He tried to show his students how to think independently and clearly. He was always very tolerant towards those of different standpoints. He wrote:

> I have always considered independence of thought as well as an appropriate method and pure love for truth the most reliable guarantee of success in scientific work. [...] Since it is my intention above all to show students who devote themselves to philosophy the right way, while allowing them to find their own way even if it is entirely inconsistent with my vision. (Twardowski 1926: 30)

This self-description is fully confirmed by his students, who stressed that Twardowski never imposed any particular views on them. He only placed emphasis on a strict formulation of theses, detailed analysis of problems and sufficient justification of the results.

The second explanation for the diversity of philosophical standpoints within the LWS is the effect of the theoretical programme realised by Twardowski. Today his approach would be called interdisciplinary. Łukasiewicz ironically reminisced about Twardowski's seminar:

> All the time one talked about whether conviction is a psychical phenomenon of a specific kind or if it is a complex of ideas; all the time one spoke about images, presentations, concepts, their content and object, and one never knew whether these analyses belonged to psychology, to logic, or to grammar. (Łukasiewicz 2013: 65)

This is a very apt description of what is done nowadays in cognitive studies and other paradigmatic interdisciplinary research. Only the scope of disciplines and tools is different now than in Twardowski's times.

However, the majority of Twardowski's students became well-disciplined thinkers: they specialised themselves in particular directions. The intensive development of their disciplines contributed to their specialisation. The most significant example is the opposition between psychologists from Witwicki's group and Łukasiewicz's group of logicians. The differences became so deep that there were few points of contact between representatives of Twardowski's school of different orientations.

The inner tension and animosities which occurred within the LWS are presented by Witwicki in his letter to Twardowski dated January 11, 1920. The passage quoted below illustrates very well the Warsaw philosophical community. Witwicki is writing in reply to Twardowski's letter in which his former teacher encouraged the student to prepare an article about the weaknesses of Warsaw logicians.

> I do not know how much I really hurt when Łukasiewicz and Leśniewski talk about "sentences", just like they would talk about grammar, and about words instead of things, objects, facts, assertions, negations, cognitions, the subjective world and the knowing subject, their mutual relationships – words, words, and more words. These "sentences" stripped clean of any traces of belief– become for me mere combinations of more or less articulated mutter. […] I am interested in sentences only with regard to the thoughts they express, convictions they reveal. Otherwise, sentences and names are objects of acoustics or phonology rather than philosophy. I do not say: "convictions, their sources, genesis and descriptions". No. This is what psychologists would examine empirically. It seems to me, however, that there are connections between words on the one hand, and between the meaning of words, things and cognitions on the other. And there are, it seems to me, difficulties which are caused by expressions, and there are other problems which are caused by things and cognitions. I cannot stop believing in the existence of properties just because Leśniewski makes some tricks with the word "property", and because of these tricks Kotarbiński states, in all seriousness, that he does

not believe in properties, does not accept properties; what he is saying, in fact, is that one sheet of white paper has nothing in common with another sheet of white paper.

In the end, Witwicki never published the article. One year later, however, the famous text 'Symbolomania and Pragmatophobia' (Twardowski 1921) appeared, in which Witwicki's thoughts are expressed by Twardowski in a more elegant and subtle form.

Witwicki's opinion about Łukasiewicz and his colleagues may be compared to Łukasiewicz's harsh comments about Witwicki found in his 'Diary':

Witwicki was my competitor in Twardowski's seminar. Twardowski tried to treat us equally. It was not easy, because we worked in different fields and in different directions. Witwicki has no talent or passion for creative, scientific work. He has written a psychology textbook, published popular works in philosophy and the history of art, and translated Plato. (Łukasiewicz 2013: 63)

Łukasiewicz, the leader of the logic wing of the LWS, presented his own programme as follows:

Philosophers, even the greatest ones, in creating their systems do not use the scientific method. The concepts they use are usually unclear and ambiguous, their statements usually neither understandable and nor justified, their reasoning almost always erroneous. All these philosophical systems have probably some significance in the history of human thought, they sometimes are of some aesthetic or ethical value, they sometimes even contain some correct statements, based on intuition; however they are of no scientific value. [...] Logic, created by mathematicians, by establishing a new measure of scientific strictness, much higher than any earlier measures of exactness, opened our eyes to the misery of philosophical speculations. [...] Future scientific philosophy must begin its reconstruction from the very beginning, from the foundations. [...] The most appropriate method which should be applied to this purpose seems to be [...] the method of mathematical logic: the deductive, axiomatic method. (Łukasiewicz 1928: 42)

However, the postulate of the axiomatisation of philosophy was accompanied, by Łukasiewicz, by a significant comment:

> The obtained results should be constantly monitored and compared with intuitive and experiential data as well as with the results of other sciences, especially the natural sciences. In case of discrepancies, the system should be corrected by formulating new axioms and creating new primary concepts. One should always take care to maintain contact with reality in order not to create mythological entities of the type of Platonic ideas or Kant's 'things in themselves', but rather, to understand the essence and construction of this real world which we live and act in, and which we would somehow like to transform into a better and more perfect one. (Łukasiewicz 1928: 42)

This reads like a reply to Witwicki's objections expressed in the letter to Twardowski—objections which he also expressed many times directly to the coryphaei of the Warsaw School of Logic.

Let these examples of interpersonal relationships between Twardowski and his students, and of the intertextual relationships between elements of their work, serve as a contribution to the picture of the LWS.

References

Ajdukiewicz, K. (1937). Kierunki i prądy filozofii współczesnej [Trends and currents of contemporary philosophy]. In K. Ajdukiewicz, *Język i poznanie* [*Language and Cognition*] (Vol. I, pp. 249–263). Warszawa: PWN.

Ajdukiewicz, K. (1951). On the notion of existence. In K. Ajdukiewicz, *The scientific world-perspective and other essays* (pp. 209–221). Dordrecht: Reidel.

Ajdukiewicz, K. (1959). Pozanaukowa działalność Kazimierza Twardowskiego (działalność dydaktyczna i organizacyjna) [Non-scientific activities of Kazimierz Twardowski (Didactic and organizational activities)]. *Ruch Filozoficzny, 19*(1–2), 29–35.

Baley, S. (1936/1937). Prof. Kazimierzowi Twardowskiemu w 70-tą rocznicę urodzin [To professor Kazimierz Twardowski on his 70th birthday]. *Polskie Archiwum Psychologii, IX*(2), 65–67.

Betti, A. (2016). Kazimierz Twardowski. In N. Edward (Ed.), *Stanford encyclopedia of philosophy*. https://plato.stanford.edu.

Czeżowski, T. (1925). Teoria pojęć Kazimierza Twardowskiego [The theory of concepts of Kazimierz Twardowski]. *Przegląd Filozoficzny, XXVIII*, 106–110.

Czeżowski, T. (1940). On happiness. In T. Czeżowski (2000), *Knowledge, science and values* (pp. 159–162). Amsterdam: Rodopi.

Czeżowski T. (1948). W dziesięciolecie śmierci Kazimierza Twardowskiego [On the 10th anniversary of the death of Kazimierz Twardowski]. In T. Czeżowski, *Odczyty filozoficzne* [Philosophical lectures] (pp. 9–16). Toruń: TNwT.

Czeżowski, T. (1953a). *On the method of analytic description*. In T. Czeżowski (2000), *Knowledge, science and values* (pp. 42–51). Amsterdam: Rodopi.

Czeżowski, T. (1953b). *On the problem of induction*. In T. Czeżowski (2000), *Knowledge, science and values* (pp. 52–59). Amsterdam: Rodopi.

Czeżowski, T. (1954). Zarzut niejasności (przyczynek do teorii dyskusji) [The charge of ambiguity (Contribution to the theory of discussion)]. In T. Czeżowski (1958), *Odczyty filozoficzne* [Philosophical lectures] (pp. 251–259). Toruń: TNwT.

Czeżowski, T. (1963). The classification of reasonong and its consequences in the theory of science. In T. Czeżowski (2000), *Knowledge, Science, and Values* (pp. 119–133). Amsterdam: Rodopi.

Dąmbska, I. (2015). *Knowledge, language and silence*. Leiden-Boston: Brill-Rodopi.

Kotarbiński, T. (1929). *Elementy teorii poznania, logiki formalnej i metodologii nauk* [Elements of the theory of cognition, formal logic and methodology of sciences]. Translated into English as: T. Kotarbiński (1966) [Gnosiology. The scientific approach to the theory of knowledge]. Oxford and Wrocław: Pergamon Press-Ossolineum.

Kotarbiński, T. (1936). Kazimierz Twardowski. In T. Kotarbiński (1979), *Szkice z historiifilozofii i logiki* [Sketches in the history of philosophy and logic] (pp. 257–264). Warszawa: PWN.

Kotarbiński, T. (1951). Kurs logiki dla prawników [A course of logic for lawyers]. In T. Kotarbiński (1993), *Ontologia, teoria poznania i metodologia nauk* (pp. 273–433). Wrocław: Ossolineum.

Kotarbiński, T. (1952). Odpowiedź na tekst Bronisława Baczki o poglądach społeczno-politycznych Kotarbińskiego [A reply to Bronisław Baczko's text on Kotarbiński's socio-political views]. In T. Kotarbiński (1993), *Ontologia, teoria poznania i metodologia nauk* (pp. 170–182). Wrocław: Ossolineum.

Kotarbiński, T. (1967). Spostrzeżenia w sprawie sposobów urabiania postawy i uzdolnień młodych pracowników naukowych [Some reflections on the methods of moulding the attitude and abilities of young scholars]. In A. Matejko (Ed.) (1967), *Kierowanie pracą zespołową w nauce* [Managing team work in science] (pp. 75–86). Warszawa: PWN.

Kotarbiński, T. (1973). Dobra robota w filozofii [Good work in philosophy]. In. T. Kotarbińśki (1986), *Drogi dociekań własnych: fragment filozoficzne* [Routes of my own investigations: Philosophical fragments] (pp. 144–145). Warszawa: PWN.

Lam, S. (1968). *Życie wśród wielu* [Life among many]. Warszawa: PIW.

Łempicki, S. (1938). Rola Kazimierza Twardowskiego w uniwersytecie i społeczeństwie [The role of Kazimierz Twardowski in the university and society]. In Kazimierz Twardowski (Ed.), *Nauczyciel, uczony, obywatel* (pp. 31–49). Lwów: PTF.

Leśniewski, S. (1992). *Collected works* (Vol. I–II). Warszawa and Dordrecht: PWN-Kluwer.

Łukasiewicz, J. (1904). O stosunkach logicznych [On logical relations]. In J. Łukasiewicz (1998), *Logika i metafizyka* [Logic and metaphysics]. *Miscellanea* (p. 97). Warszawa: WFiS UW.

Łukasiewicz, J. (1911). O rodzajach rozumowania. Wstęp do teorii stosunków [On types of reasoning. Introduction to the theory of relations]. In J. Łukasiewicz (1998), *Logika i metafizyka* [Logic and metaphysics]. *Miscellanea* (p. 232). Warszawa: WFiS UW.

Łukasiewicz, J. (1912). O twórczości w nauce [On creativity in science]. In: J. Łukasiewicz (1998), *Logika i metafizyka* [Logic and metaphysics]. *Miscellanea* (pp. 9–33). Warszawa: WFiS UW.

Łukasiewicz, J. (1928). O metodę w filozofii. [For the Method of Philosophy]. In J. Łukasiewicz (1998), *Logika i metafizyka* [Logic and Metaphysics]. *Miscellanea* (pp. 41–42). Warszawa: WFiS UW.

Łukasiewicz, J. (1929). O znaczeniu i potrzebach logiki matematycznej [On the importance and needs of mathematical logic]. In J. Łukasiewicz (1998), *Logika i metafizyka* [Logic and metaphysics]. *Miscellanea.* (pp. 424–437). Warszawa: WFiS UW.

Łukasiewicz, J. (1998). *Logika i metafizyka* [Logic and metaphysics]. *Miscellanea.* Warszawa: WFiS UW.

Łukasiewicz, J. (2013). *Pamiętnik* [Diary]. Warszawa: Semper.

Perzanowski, J. (Ed.). (2009). *Rozum – serce – smak. Pamięci Profesor Izydory Dąmbskiej* [Mind—Heart—Taste. In memory of professor Izydora Dąmbska]. Kraków: WAM.

Smith, B. (1989). Kasimir Twardowski: An essay on the borderlines of ontology, psychology and logic. In K. Szaniawski (Ed.), *The vienna circle and the Lvov-Warsaw School* (pp. 313–373). Dordrecht: Kluwer.
Szaniawski, K. (1983). Filozofia w oczach racjonalisty. Słowo wstępne [Philosophy in a rationalist's eyes. An introduction]. In K. Ajdukiewicz (1983), *Zagadnienia i kierunki filozofii* [Problems and trends in philosophy]. Warszawa: Czytelnik.
Twardowski, K. (1894). *Zur Lehre vom Inhalt und Gegenstand der Vorstellungen.* Wien: Hölder.
Twardowski, K. (1899). Pesymism i optymism. In K. Twardowski (2014), *Myśl, mowa i czyn* [Thought, speech and action] (pp. 329–332). Warszawa: Semper.
Twardowski, K. (1901). *Zasadnicze pojęcia dydaktyki i logiki* [The basic concepts of didactics and logic]. Lwów: Towarzystwo Pedagogiczne.
Twardowski, K. (1903). The essence of concepts. In K. Twardowski (1999), *On actions, products and other topics in philosophy* (pp. 73–97). Amsterdam: Rodopi.
Twardowski, K. (1904/1905). Ethics, criminal law and the problem of free will. In K. Twardowski (1999), *On prejudices, judgments and other topics in philosophy* (pp. 287–321). Amsterdam: Rodopi.
Twardowski, K. (1912). Actions and products. In K. Twardowski (1999), *On actions, products and other topics in philosophy* (pp: 217–240). Amsterdam: Rodopi.
Twardowski, K. (1913). O psychologii, jej przedmiocie, zadaniach, metodzie, stosunku do innych nauk i o jej rozwoju [On psychology, its subject, method, relation to other sciences and on its development]. In K. Twardowski (2014), *Myśl, mowa i czyn* [Thought, speech and action] (pp. 241–291). Warszawa: Semper.
Twardowski, K. (1914). *O pojęciu rozumowania* [On the notion of reasoning] (unpublished).
Twardowski, K. (1920). On clear and unclear philosophical style. In K. Twardowski (1999), (pp. 257–259). Amsterdam: Rodopi.
Twardowski, K. (1921). Symbolomania and pragmatophobia. In K. Twardowski (1999), *On actions, products and other topics in philosophy* (pp. 261–270). Amsterdam: Rodopi.
Twardowski, K. (1923/1924). On ethical scepticism. In K. Twardowski (2014), *Myśl, mowa i czyn* [Thought, speech and action] (pp. 237–286). Warszawa: Semper.

Twardowski, K. (1926). A self-portrait. In K. Twardowski (1999), *On prejudices, judgments and other topics in philosophy* (pp. 17–31). Amsterdam: Rodopi.

Twardowski, K. (1927). Z logiki przymiotników [On the logic of adjectives]. In K. Twardowski (1999), *On actions, products and other topics in philosophy* (pp. 141–143). Amsterdam: Rodopi.

Twardowski, K. (1931). Przemówienie na uroczystości wręczenia medalu pamiątkowego [Address on the award of a commemorative medal]. In K. Twardowski (2014), *Myśl, mowa i czyn* [Thought, speech and action] (Vol. 2, pp. 452–455). Warszawa: Semper.

Twardowski, K. (1934). *Teoria badań indukcyjnych* [The theory of inductive studies] (unpublished).

Twardowski, K. (1965). *Wybrane pisma filozoficzne* [Selected philosophical writings]. Warszawa: PWN.

Twardowski, K. (1999). *On actions, products and other topics in philosophy.* Amsterdam: Rodopi.

van der Schaar, M. (2015). *Kazimierz Twardowski: A grammar for philosophy.* Leiden: Brill.

Witwicki, W. (1900). Analiza psychologiczna ambicji [Psychological analysis of ambition]. *Przegląd Filozoficzny, III*(4), 26–49.

Witwicki, W. (1904). Analiza psychologiczna objawów woli [Psychological analysis of manifestations of the will]. *Archiwum Naukowe, II,* 1–27.

Witwicki, W. (1905). Analiza psychologiczna pragnień jako podstawa etyki [Psychological analysis of desires as the basis of ethics]. *Muzeum, 21,* 1026–1038.

Witwicki, W. (1912). W sprawie przedmiotu i podziału psychologii [On the Subject Matter and Classification of Psychology]. In *Księga pamiątkowa ku uczczeniu 250 rocznicy założenia Uniwersytetu Lwowskiego* [Memorial book to commemorate the 250th anniversary of the founding of the Lvov University] (Vol. II, pp. 1–16). Lwów: Uniwersytet Lwowski.

Witwicki, W. (1920). Kazimierz Twardowski. *Przegląd Filozoficzny, 23,* IX–XIX.

Witwicki, W. (1925). Rozmowa z pesymistą [A conversation with a pessimist]. *NaszaPraca, IV*(9–10), 153–164.

Witwicki, W. (1931). O źródłach poznania życia uczuciowego [On the sources of knowledge in emotional life]. In *Księga pamiątkowa Polskiego Towarzystwa Filozoficznego we Lwowie* [Memorial book of the polish philosophical society in Lvov]. Lwów: PTF.

Witwicki, W. (1932). W sprawie teorii postanowień [Regarding the theory of resolutions]. *Kwartalnik Psychologiczny, II*, 492–502.

Witwicki, W. (1938). Kazimierz Twardowski. *Wiadomości Literackie, 16*(18), 1.

Witwicki, W. (1939). *La foi des éclairés*. Paris: Alcan.

Witwicki, W. (1947). Podział życia psychicznego. Uwaga, pamięć, fantazja [Classification of mental life. Attention, memory, fantasy]. *Wiadomości z psychologii, 4*, 1–20.

Witwicki, W. (1957). *Pogadanki obyczajowe* [Moral causeries]. Warszawa: PWN.

Woleński, J. (1989). *Logic and philosophy in the Lvov-Warsaw school*. Dordrecht: Kluwer.

Part II

Psychology

7

The Relationship Between Judgments and Perceptions from the Point of View of Twardowski's School

Stepan Ivanyk

1 Introduction

'Judgment' and 'perception' were undoubtedly two of the most important terms used in psychological and logical theories developed in the School founded by Kazimierz Twardowski. The issue of the mutual relationship between the correlates of these terms has been a concern for many of the School's representatives for many decades (cf. Benjamin 2007). No comprehensive study of analyses they have performed in this respect has appeared in philosophical literature so far, however. The main goal of this article is to fill this gap by pursuing the following specific objectives:

This paper has been published with the support of the National Science Centre in Krakow under the 'Sonata-5' Programme (Project name: 'The Theory of Judgment in the Lvov School of Philosophy'; Project No.: 2013/09/D/HS1/00690).

S. Ivanyk (✉)
Kazimierz Twardowski Philosophical
Society of Lviv, Lviv, Ukraine

A. Drabarek et al. (eds.), *Interdisciplinary Investigations into the Lvov-Warsaw School*, History of Analytic Philosophy,
https://doi.org/10.1007/978-3-030-24486-6_7

1. Review the evolution of views held by the School's representatives on the relationship between judgments and perceptions;
2. Summarise and classify their main findings.

2 Twardowski

The first work in which Twardowski analysed the relationship between perception and judgment was his Ph.D. thesis entitled 'Idea i percepcja. Z badań epistemologicznych nad Kartezjuszem' ('Idea and Perception—an Epistemological Investigation of Descartes', 1892). The goal of his dissertation was to determine the relationship between perception and idea in René Descartes, in view of the fact that many researchers (e.g. Julius Koch and Paul Natorp) considered Descartes' expressions to have the same meaning. In fact, in the terminology characteristic of Twardowski's philosophy, the term 'idea' used by Descartes can be appropriately 'converted' to 'representation'. Thus, Twardowski's goal was to analyse Descartes's views on the relationship between perception and representation. But since *clara et distincta perceptio* (clear and distinct perception) in the theory of Descartes acted as the fundamental criterion of truth (i.e. true judgment), the relationship between perception and representation could not be analysed without taking into account the relationship between perception and judgment.

Therefore, Twardowski devoted much of his attention to analysing the latter relationship. In the 'Preface by the Translator' to the Polish translation of Twardowski's work, Elżbieta Paczkowska explicitly mentioned that 'of peculiar importance here is Twardowski's approach to perception, especially his attitude to judgment' (cf. Paczkowska-Łagowska, 1976). Let us take a look at Twardowski's analyses and try to explicate his position concerning the relationship between perception and judgment.

As is well known, Descartes distinguished three basic classes of psychological phenomena:

1. Representations;
2. Judgments;
3. Affects.

Since perception belongs neither to the class of representations,[1] nor to the class of affects (I am disregarding details of Twardowski's analyses as irrelevant to our discussion), the question arises whether Descartes included perceptions in the class of judgments.

While reconstructing Descartes's theory of judgment, Twardowski differentiates its parts in the following way:

1. Perception (*ratio*, precondition of the judgment);
2. Representation (*materia*, subject matter of the judgment);
3. Decision of will (consent to making the judgment);
4. Affirmation or rejection (the form of judgment).

Descartes also clearly distinguishes two types of perception—*perception sensu* and *perception ab intellectu*—which in Twardowski's terminology can be converted to 'internal perception' (or 'introspection') and 'external perception' ('extrospection'). For Descartes, internal perception is a special case, because we are always dealing here with clarity and distinctness of the psychological act, which in a way guarantees the rightness of the judgment. The link (3) in the chain of elements required to form a judgment emerges automatically, so to speak, because clear and demonstrable perception determines the will and unconditionally induces us to make positive or negative judgments.

One way or another (irrespective of whether we are dealing with introspection or extrospection)—it is clear that Descartes did not equate perception and judgment. According to Descartes, perception is not judgment because:

1. it is its precondition (necessary, though not sufficient) which induces the will to make a judgment;
2. perception as such does not involve a decision of the will, without which judgment is impossible;
3. perception does not have a specific attribute of judgment which is its truth value (true or false).

---

[1]In this point, Twardowski opposed the interpretation suggested by his Master,Franz Brentano, according to whom it is possible to include what Descartes called perception among representations (cf. Twardowski 1892).

Therefore, Twardowski concluded, in Descartes' philosophical system, perception cannot be included in any of the three basic classes of psychological phenomena. Twardowski asserted this fact with some regret. He wrote, for example, 'The fact that it is not possible to classify them [perceptions] as belonging to any of these three classes can only be explained on the assumption that Descartes did not recognise the essence of perception' (Twardowski 1892/1976: 328).

One may conjecture that the fact perception could not be included in any of the three classes of psychological phenomena caused some discomfort to Twardowski, because he had already developed his own view on the subject, different from that held by Descartes. What was this view? We have all reason to believe that, at least when writing his doctoral dissertation, Twardowski was inclined to include perception in the class of judgments. Here is some suggestive evidence:

1. Analysing the role of clear and distinct perception in Descartes's theory, Twardowski revealed his own position by stating that 'Descartes's criterion [of the truthfulness of judgments] is quite rightly thought-out, but it is only to consider the precondition of judgment, as Descartes calls the perception, as judgment itself' (Twardowski 1892/1976: 335). Moreover, Twardowski even elaborated a little on this suggestion, which shows us what mistake Descartes made in his treatment of the essence of internal perception. Twardowski pointed out that it is necessary to distinguish between two types of internal perception, depending on what objects it is directed at: (a) psychological phenomena (acts of presenting, judging, feeling, wanting) and (b) immanent objects (the contents of presenting, judging, feeling, wanting). In Twardowski's opinion, Descartes (justifying *cogito ergo sum*) took into account only the first type of internal perception, while entirely disregarding the second one.
   On the other hand, if one accepts the condition that objects in the second group of perceptions are abstract representations (concepts), the obviousness of judgments made on the basis of these internal perceptions will be maintained. This way, Twardowski proposed a classification of obvious judgments: (a) perceptual judgments—in the first group of internal perceptions and (b) analytical judgments—in the second group of internal perceptions.

2. When writing his doctoral dissertation, Twardowski was a supporter of Franz Brentano's idiogenic theory of judgments. Brentano (in opposition to Aristotle's allogenetic theory of judgments) stated that the form of judgment could be different from that Aristotle talked about (i.e. the form of categorical judgments comparing scopes of two or more representations). Brentano's theory affected his view of the relationship between perception and judgment. Namely, Brentano's perception is a kind of such judgment:

> By "judgment" we mean, in accordance with common philosophical usage, acceptance (as true) or rejection (as false) [...] Such acceptance or rejection also occurs in cases in which many people would not use the term "judgment," as, for example, in the perception of mental acts and in remembering. That predication is not the essence of every judgment emerges quite clearly from the fact that all perceptions are judgements, whether they are instances of knowledge or just mistaken affirmations. (Brentano 2009: 153–162)

The evidence for the fact that Twardowski understood the relationship between perception and judgment similarly to Brentano can be found in his dissertation, which contains a strong rejection of Natorp's assertion that the examples of judgments provided by Descartes should only be treated as combinations of representations. In his interpretation of Descartes' reply to the question about what is affirmed or denied in a judgment, Twardowski wrote:

> One could consider various types of representations containing a set of many attributes, e.g. the nature of a triangle, square or some other figure; similarly, the nature of the mind or body, but first of all, the nature of God, the most perfect being. We would then notice that everything that is recognised as included in them can be stated as true. (Twardowski 1892/1976: 325)

Here, as Twardowski pointed out, we are not dealing with a combination of representations, but rather with an analysis of the content of one representation (as in the judgment: 'Matter is spatial').

The view that the relationship between judgment and perception in fact consists in the latter being a special case of the former significantly evolved in Twardowski's later works—'O treści i przedmiocie przedstawień' ('On the Content and Object of Representations', 1894, and particularly in 'Images and Concepts', 1898).

In 'Images and Concepts', the relationship between judgments and perceptions is mentioned in the context of considering whether it is possible to regard images as recreations of perceptions. Agreeing with Hippolyte Taine and Alois Höfler, Twardowski wrote that perception is a complex psychological act which includes judgment, among other things: 'It is beyond doubt that perceptions include judgments' (Twardowski 1965: 124). Therefore, images should not be considered as recreations of perceptions, because judgments, included in perceptions, are not recreated in images.

Judgment which is 'included' in perception has a thetical function: it is responsible for the 'localisation and objectification' of groups of impressions, which are also included in perception. This is an affirmative judgment stating that something we perceive is located in a particular place, or that it is a particular object. In Twardowski's work, we have some clarification as to which type of judgment we are dealing with in perceptions. Let us try to define the basic forms of these judgments:

(1) 'Localising' judgments:
'A perceived object X at the time of perception T is in a particular place P'.
In Twardowski's example: A person who sees another person sitting in front of them at a table makes the judgment that this person is in this place at this moment.
(2) 'Objectifying' judgments:
'A perceived object X is a particular object'.

This expression is quite vague. It might seem that this is a case of judgments in which perceived objects are classified into classes (recognised as types) of other objects. However, the example provided by Twardowski shows that this is precisely about existential judgments. He wrote: 'Who sees a cloud, for example, and is at the same time convinced of its existence' (Twardowski 1965: 124).

Therefore, the basic form of 'objectifying' judgments should recognise the following:

> 'A perceived object X exists'.

The proof that we are dealing with judgments here consists, according to Twardowski, in the fact that perceptions have the specific attribute of judgments—the truth value. They can be true or false (the latter in the case of sensual illusions). As can be seen, Twardowski borrowed this argumentation from Brentano.

Let us sum up Twardowski's stance as follows:

1. Perceptions are complex psychological acts which include judgments. The proof of this is, among others, the fact that perceptions, like judgments, can be right (in the case of normally functioning senses) or wrong (in the case of sensory illusions);
2. The inclusion of judgments determines the fundamental difference between perception and representation: perceiving an object is always accompanied by a conviction that it exists; presenting an object, however, does not have to be associated with the belief that it exists (or does not exist).
3. In view of the above, we can see that in comparison with 'Idea and Perception', Twardowski's stance expressed in 'Images and Concepts' (a) has been slightly modified (perceptions as such are not judgments, but only include judgments) and (b) clarified (in perceptions we are dealing with so-called 'localising' and 'objectifying' judgments).

As can be seen, the relationship between judgment and perception according to Twardowski already consists in the fact that the former is 'included' in the latter. What did Twardowski mean by 'included', however? It seems that this term is rather vague, and unfortunately Twardowski did not explain this relationship. He could have meant very different things, for example, that (1) judgment is an element of perception; (2) judgment is a kind of perception; and (3) judgment always accompanies perception, etc.

In his articles 'O kłopotach związanych z tetycznością' ('On Problems Related to Theticity', 1999) and 'W sprawie poczucia realności świata' ('On the Sense of the World's Reality', 2000), Adam Olech pointed out some other, no less serious shortcomings of Twardowski's view discussed above (Olech 2000: 334–360). Comparing views on the essence of the thetical moment (i.e. conviction about the reality of perceived objects, or the 'sense of the world's reality') developed in Twardowski's school with those originating from the phenomenological tradition, the author states that phenomenological analyses (in particular those by Edmund Husserl and Roman Ingarden) were much more advanced and accurate in this respect than the analyses of Twardowski and his students.

Olech noticed that Twardowski and his students (following the footsteps of Brentano) believed that the thetical moment was a judgment, which being a simple and independent psychological act itself was a component of the complex act of perception. Husserl and Ingarden, on the other hand, claimed that perception was a simple and independent act, and that the thetical moment included in it was its dependent component. According to Husserl and Ingarden, the thetical moment lies in perception, rather than in the existential judgment.

The main cause of this difference in Olech's opinion consists in the fact that phenomenologists embraced a more developed ontology, including both objects and states of affairs. They acknowledged that in perception of 'being' of object was recognised (it is not a judgment about the existence of this object), and only later, in the judgment built on this perception—existence of the object. Consequently, for phenomenologists the thetical act is not a judgment, but perception. Olech concluded that Twardowski's school lacked an ontology of the state of affairs, which resulted in attributing the thetical moment above all to judgments, and only through them—to perceptions.

Thus, the question why the judgment included in perception (so-called perceptual judgment) has these particular constitutive features was, in Olech's opinion, almost entirely disregarded by Twardowski and his students. Here are some of Olech's very interesting comments in this respect:

> The thetical moment included in the act of perception (Setzungsmoment), which in Husserl and Ingarden's opinion is not an independent act [...], is described by Twardowski as an affirmative(recognising) judgment and classified as an autonomous act (to use Ingarden's terminology). It relies – in Twardowski's opinion – on the reference of "impressions experienced in a visual representation to a particular object and place" [...] And that is all, nothing more is said on the subject. Brentano was similarly laconic [...] Neither in the works of Brentano nor Twardowski is there a detailed analysis of the act of perception. [...] The metaphysical realism of Twardowski and his students has its natural source in their natural attitude to the world. It was not, as far as I know, subjected to critical inspection in Twardowski's school. We might say that the school's realistic attitude ultimately derives its realistic value from its uncritical natural attitude. (Olech 1999: 212–216)

The above passage contains two implicit charges against the entire Lvov school of philosophy:

1. the charge of scarcity and poor quality of the analyses of perception;
2. the charge of uncritical natural attitude.

Let us add to one more:

3. the charge of inaccurately defined concept of 'inclusion' used to describe the relationship between judgment and perception.

Let us now ask whether these three charges are indeed valid in relation not only to Twardowski, but also to his school as a whole.

However, before we take a look at the views of Twardowski's students on the relationship between judgment and perception, it should be noted that the first charge seems to be too exaggerated even in relation to Twardowski himself. In his works, Olech only refers to two of Twardowski's works—'On the Content and Object of Representations' and 'Images and Concepts'. Admittedly, Twardowski devoted very little attention to perceptions in his works, or none at all. However, it seems that Olech did not take into account one of his earlier works namely, his doctoral dissertation of 1892 (in which Twardowski said a lot about perception), mentioned at the very beginning of this chapter.

3 Students of Twardowski

The first of Twardowski's students to express an opinion (however laconically) about the relation between judgments and perceptions was the Ukrainian-Polish psychologist Stefan Baley. In his handbook 'Zarys psychologii' ('An Outline of Psychology', 1922, he wrote:

> Apart from representation of a perceived object, perception also includes the awareness that we are dealing with a real object and not with a product of fantasy. This awareness is also a judgment about its being. (Baley 1922/2002: 328)

A statement found in Baley's later handbook entitled 'Zarys psychologii w związku z rozwojem psychiki dziecka' ('An Outline of Psychology in Relation to the Development of a Child's Psyche', 1935) indicates that he was inclined to understand the 'inclusion' of judgment in perception as accompaniment, co-occurrence: 'An analysis of the act of perception shows that apart from images, certain judgments are included in it as well. Perception is accompanied by conviction about the reality of what I perceive' (Baley 1922/2002: 235).

Another of Twardowski's student, Władysław Witwicki, in his handbook 'Psychologia' ('Psychology', 1925) admitted that perception was a complex psychological phenomenon. According to Witwicki, it consists of the following elements:

1. representation of an object—perceptual image;
2. realising judgment—judgment about the object's existence;
3. classifying judgments (at least one)—judgments that classify the object into a certain class of objects.

There are certain correlations between perceptions and the features of realising and classifying judgments which make it up:

1. The more extensive the class of objects to which a perceived object is referred, the less accurate and clear the perception is;
2. Depending on whether the realising judgment or some of the classifying judgments are true or false, perception is true or false as well. False perception based on a false realising judgment is a hallucination (e.g. seeing a dead person); false perception based on a false classifying judgment is an illusion (e.g. when parallel lines are perceived as divergent or convergent).

Special attention should be paid to so-called conscious illusions. They occur when new classifying judgments correct erroneous classifying judgments which have caused a primary illusion, so that while we still experience the illusion, it is now experienced 'consciously' (e.g. when we realize that lines which we erroneously perceived as divergent are in fact parallel, and yet we still perceive them as divergent). In Witwicki's opinion, conscious illusions should be investigated more thoroughly in order to determine:

1. whether they consist in a simultaneous appearance of two contradictory convictions in our awareness;
2. whether they consist in a non-simultaneous appearance of two contradictory convictions in our awareness;
3. whether one of the two contradictory convictions is not actually a conviction, but a so-called 'quasi-conviction', i.e. fiction.

In the 1930s, Witwicki's view on the structure of perception and the place of judgments in this structure was developed and elaborated by his students in Warsaw. For example, in 'Dwa rodzaje spostrzeżeń i ich struktura' ('Two Kinds of Perception and Their Structure', 1935), Maria Adler drew attention to the difference between the content of realising judgments included in external ('objective') and internal ('subjective') perceptions. The difference results in that in external perceptions, the object of perceptual image is located externally, and in the case of internal perceptions, it is located internally (e.g. we locate our rheumatic pain in some of our muscles).

In his article 'Spostrzeżenia i przypomnienia' ('Perceptions and Reminders', 1933), Tadeusz Czeżowski criticised Witwicki's theory of perceptions. According to Czeżowski, the weak point of this theory consists in the inclusion of perception in the structure of so-called classifying judgments. Czeżowski argued as follows:

1. Accepting the classifying judgment as an element of perception results in the acceptance, next to a representation of the perceived object, also other representations as forming part of perception, such as: representation of a class, representation of an object similar to the object of perception, etc. However, introspection does not suggest any of these representations that are present in the act of perception.
2. The assumption that some other judgments should be included in perception, in addition to the realising judgment, is unnecessary. The function Witwicki assigned to classifying judgments is that of presenting the perceived object. This way it is possible to better explain the essence of illusions, without involving the classifying judgment. An illusion, according to Czeżowski, arises in when the perceptual image does not provide all information necessary to build a comprehensive representation of the object, and the missing fragments are supplemented through reconstruction. An object is then presented having a different quality than the object actually perceived, but the realising judgment affirms its existence.

One of Twardowski's less known student, the Ukrainian philosopher Stefan Oleksiuk discussed the relationship between perception and judgment in his Ph.D. thesis entitled 'O tzw. sądzie spostrzeżeniowym. Studium z psychologii poznania' ('On the So-Called Perceptual Judgment. A Study in the Psychology of Cognition', 1932), devoted to the essence of judgment included in perception. The subject of Oleksiuk's analysis was the 'so-called perceptual judgment'. According to Oleksiuk, it is a judgment which is included in the structure of external perception, and which determines precisely that we attribute actual existence to perceived objects of the external world. Oleksiuk also listed some negative attributes of perceptual judgment to help understand its essence and place in the structure of perception. Thus, the main characteristics of perceptual judgment are as follows:

1. It is not a judgment made on the basis of perception (because it is something 'embedded' in perception);
2. It is not a judgment referring to the existence of an object, appearing beside judgments about its remaining properties: perceptual judgment also refers to all the other properties of the perceived object;
3. It is not an experience concerning both the external and the internal world: it is only concerned with the external reality.

The main problem associated with perceptual judgment is as follows: How is it possible that while dealing only with aspectual and changeable representations of an object, perceptual judgment makes us feel the reality of a perceived object as comprehensive, 'identical in itself'? The issue addressed by Oleksiuk is a version of one of the main concerns of the theory of cognition: the problem of the constitution of an object in the changeable stream of phenomena, or the so-called problem of psychological transcendence.

In an attempt to confront the problem of psychological transcendence, Oleksiuk undertook to answer the question whether it is possible to solve this problem respecting the principle which could be called 'the principle of phenomenality'. This principle considers representation as our basic attitude to the objects of external world—as a necessary psychological basis for 'founded' experiences (i.e. those built on representations). If in the experience of reality we are 'doomed' to phenomena, then we must agree that we are referring to objects of the external world only through phenomena. However, phenomena are always aspectual (variable, one-sided), whereas objects in external perception appear to us as 'identical in itself', that is, as we would say today, gen-identical. The result of analyses performed by Oleksiuk is negative: having rejected all positions which try to explain the essence of perceptual judgment while respecting the principle of phenomenality,[2] Oleksiuk concludes that the

[2]These positions were the following: (1) the object of perceptual judgment is a thing as such, taken directly (the radical realism of T. Kotarbiński); (2) the object of perceptual judgment is a

sense of reality of particular objects in their external perception cannot be explained in terms of representations. According to Oleksiuk, this is demonstrated by the fact that the principle of 'phenomenality' is not as obvious and omnipresent as it appeared to Brentano and his students, as well as others, and the problem of psychological transcendence should (and could, Oleksiuk believed) be solved without it.

Let us notice that the conclusion Oleksiuk makes about insufficiency of the principle of phenomenality in solving the problem of psychological transcendence is surprisingly similar to the following comment about perception made by Olech:

> Contrary to Brentano's and Twardowski's view, the act of perception in Husserl's and Ingarden's views does not consist of a thetically neutral representation and a thetic, i.e. asserting existence, act of thinking, referring to what was previously only presented, and therefore does not consist of representation and judgment, but is a single indivisible whole. (Olech 2000: 357)

However, Oleksiuk admitted that despite the fact that a judgment about the actual existence of an object is evidently present in perception, the way by which we arrive at this judgment is yet to be discovered.

On a different level, Henryk Mehlberg and Kazimierz Ajdukiewicz explored the question of relationships between judgments and perceptions. They were mainly interested in whether perception can provide sufficient justification for the truthfulness of perceptual judgment. In his lecture delivered during a meeting of the Polish Philosophical Society in Lvov, entitled 'The Issue of Objectivity and Adequacy of

phenomenon as such, taken directly; (3) the object of perceptual judgment is a thing identified with a phenomenon (naive realism); (4) the object of perceptual judgment is a thing which we arrive at by way of a critical analysis of the relevant phenomena (the critical realism of O. Külpe) (5) the object of perceptual judgment is a thing present in the phenomenon as in its symbol, image (the theory of signs); (6) the object of perceptual judgment is a thing thought through, intentionally embodied in or by a phenomenon (the theory of T. Lipps); (7) the object of perceptual judgment is a thing constituted in the stream of phenomena by the unity of intention or consciousness (E. Husserl); and (8) the object of perceptual judgment is a thing present in the perceptual phenomenon through participation of derivative representations (the theory of perceptual assimilation developed by H. Bergson, H. Linke, H. Driesch and D. Katz).

Perceptions and Interpretation of Perceptual Propositions', (1929/1931: 206a), Mehlberg discussed one of the most serious charges against the view of non-objectivity and inadequacy of perceptions, which relies on the fact that the object of relevant perceptual judgments is feigned (i.e. perceptual judgments do not have objects (they are non-objective)). For example, a person standing on a railway track can see the rails (and can thus make the perceptual judgment: 'I see rails'). In this case, 'seeing rails' can have very different meanings, for example:

1. Seeing two lines converging in the distance;
2. Seeing two parallel non-running bars;
3. Seeing only the upper parts of rails.

In Mehlberg's opinion, the solution to the problem of the non-objectivity of perceptual judgments should go in the direction of selecting from the many possible meanings of the object of perceptual judgment the one that is intersubjective. The meaning of such inter-subjective object of perceptual judgments was discussed by Ajdukiewicz during a lecture delivered at the meeting of the Polish Philosophical Society in Lvov, entitled 'O obiektywności poznania zmysłowego' ('On the Objectivity of Sensory Cognition', 1931). According to Ajdukiewicz, the unambiguity of the object of perceptual judgments is conditioned by the semantic rules of the language in which it is expressed.

In his handbook 'Propedeutyka filozofii' ('Introduction to Philosophy', 1948), addressing the epistemological theme of the origin of knowledge, Ajdukiewicz mentioned perception (external or internal) as one of the five 'sources' of judgments,[3] and called judgments so formed perceptual:

> The perception of an object prompts us to make a number of judgments about it, in the sense that perception constitutes a guarantee of the truthfulness of these judgments. These judgments are made simultaneously with perception, and if we were asked about the grounds on which we

[3]Among the other four 'sources' of judgments, Ajdukiewicz listed memory, inference, obviousness and confidence in other people's words.

make them, we would answer that we make them because we can see (hear, feel, etc.) what they assert. Such judgments are called perceptual. (Ajdukiewicz 1948: 10)

However, one and the same perception can be a source of not one, but many different judgments. For example, when seeing alight in a friend's window, we can make the judgment: 'A light is on in my friend's room' or another judgment: 'My friend is at home'. In the former judgment, perception performs the role of direct justification, in the latter—that of indirect justification. Ajdukiewicz also observed that both judgments, whether based on perception directly or indirectly, can be false.

At this point, two remarks arise:

1. The relationship between judgments and perceptions was understood by Ajdukiewicz (as well as Mehlberg) in an entirely different way than by other representatives of the School: judgment is no longer a component, a building block of perception, but stands with it in a causal relationship (judgment is the result of perception);
2. It seems that there is some imperfection in the above classification of perceptual judgments performed by Ajdukiewicz (as directly and indirectly based on perception). The point is that judgments indirectly based on perception are 'inferential' rather than perceptual—we make such judgments by inference, not perception. For example, the judgment: 'My friend is at home' is a conclusion derived from the premises: 'A light is on in my friend's room' and 'My friend is always in his room when the light is on'.

A confirmation of the above-mentioned remarks can be found in one of the later works by Ajdukiewicz, 'Logika pragmatyczna' ('Pragmatic Logic', 1965), where the author used a slightly different concept. Firstly, he referred to perceptions as 'inducing motives' to make judgments (which confirms our interpretation of his understanding of the relationship between perception and judgment as that of cause and effect). Secondly, instead of 'perceptual judgment', he used the term 'protocol statement', which means judgment based directly on perception (Ajdukiewicz was most clearly conscious of a deficiency of the classification performed in the 'Introduction to Philosophy', and eliminated this deficiency in 'Pragmatic Logic').

It also seems that replacement of the term 'perceptual judgment' with 'perceptual proposition' was aimed to emphasise the role of language. The absence of the latter, in Ajdukiewicz's opinion, would simply eliminate the possibility of the motivating function of perception regarding judgment:

> No observation [perception] would be a motive inducing a person to accept a statement, i.e., a certain verbal construction, should that person not know the language in which that statement is formulated. It is only by learning a given language that we acquire the ability of responding to certain observations by accepting certain statements. When learning a language, we learn how to use expressions formulated in that language in the way which is in agreement with the usage common to those who belong to that language community. (Ajdukiewicz 1965/1974: 242)

By emphasising the role of language, Ajdukiewicz was able to solve the problem of the non-scientificity of perceptual judgments (protocol statements). According to Ajdukiewicz, the problem is that perceptual judgments as such are not scientific statements, because the way in which they are justified (directly through perception) is subjective and not intersubjective as required by science.

For example, when an astronomer observing a star by telescope makes a judgment about its movement, such perceptual judgment originally cannot be regarded as scientific. It is justified only for the astronomer, whereas for other people (not having access to the particular act of the astronomer's perception)—it is not. However, this judgment can become scientific with time. Its intersubjectivity is achieved through the following mechanism:

> They [other people] state that the observer is veracious (i.e. no conscious lie is allowed) and competent (i.e. recognising protocol statements in normal conditions, he is governed by the empirical rules of language, and in conditions other than normal, he makes an appropriate correction). It results from the competence of the observer that his protocol

> statements are true. From his veracity it results, in turn, that if he claims that a given statement is a protocol statement, then (what he says, saying that he observed that it is as this statement says) such statement is also for him a protocol statement. From these premises it results by way of deduction that it is possible to rely on what the observer says is the result of his observation. (Ajdukiewicz 1965/1974: 226)

Let us note that the above explanation provided by Ajdukiewicz is a very sophisticated way of developing Twardowski's idea. In his article 'O *naukach apriorycznych czyli racjonalnych* (*dedukcyjnych*) i *naukach aposteriorycznych czyli* empirycznych (indukcyjnych)' ('On *a priori*, i.e. rational [deductive], and *a posteriori*, i.e. empirical [inductive] sciences', 1923), Twardowski stated that *a priori* sciences use exclusively deduction from adopted assumptions (axioms), while *a posteriori* sciences require reference to perceptual judgments. He did not explain, however, what such 'reference' consists in and how it should proceed.

In her article 'Uzasadnienie zdań spostrzeżeniowych: Ajdukiewicz, McDowell, Davidson' ('Justification of perceptual statements: Ajdukiewicz, McDowell, Davidson', 2013), Adriana Schetz introduced the concept developed by Ajdukiewicz into the contemporary debate about the scientific nature of perceptual judgments. She identified two opposing stances on the issue:

1. John McDowell's stance, according to which perceptual judgments are justified by perception itself;
2. Donald Davidson's stance, according to which perceptual judgments are justified not by perception, but only by other judgments.

According to Davidson, perceptual judgments are provided with scientific nature (i.e. adequate justification) by so-called triangulation—i.e. reference to the social context (consisting in recognising the verbal reaction of other persons to perceptions of the same objects), in order to establish the meaning of one's own perceptual judgment. According to Schetz (with whom I agree), the concept proposed by Ajdukiewicz was, so to speak, prior to Davidson's stance: the Polish philosopher 'could easily agree with triangulation as understood by

Davidson' (Schetz 2013: 455). But Davidson's idea of indirect justification of perceptual judgments can be perceived as a more developed version of Ajdukiewicz's idea of the scientification of perceptual judgments through their intersubjectivisation.

4 Conclusion

Going back to the three charges against the School mentioned above, i.e. (1) inadequate analysis of perception; (2) uncritically naturalist attitude; and (3) inaccuracy of the concept of 'inclusion', it should be noted that despite differences in the views on the structure of perception held by Witwicki and Adler on the one hand, and Czeżowski on the other, it seems that their analyses are detailed enough to refute charge (1), and even more importantly, that these views (in this context, Baley's views as well) clearly specify what Twardowski's notion of 'including' judgments in perceptions consists in, and in this way, they free the School from charge (3). The analyses performed by Oleksiuk, Mehlberg and Ajdukiewicz also free the School from charge (2), as they include examples of their being aware of complications which result from an uncritically naturalist attitude towards the nature of perceptual judgments, and of attempts to overcome such complications.

The School's representatives who investigated the relationship between judgments and perceptions can be divided into two groups:

1. Those who tried to determine the place of judgments and perceptions in the overall structure of the psyche by seeking to answer the question 'Are judgments a kind of perceptions (or *vice versa*)?' and 'Are judgments an element of perceptions (or *vice versa*?)' (Twardowski, Baley, Witwicki, Adler, Czeżowski, Oleksiuk). Their views evolved from the relationship of '*genus proximum-definiendum*' (Twardowski's early analyses) to that of co-existence (Baley) or a 'part-to-whole' relationship (Witwicki, Czeżowski, Adler) between judgment and perception.
2. Those who tried to determine the justifying power of perceptions for the truthfulness of judgments related to them (Mehlberg,

Ajdukiewicz). The result of research carried out by these representatives of the School was rejection of the possibility of a logical justification of judgments by relevant perceptions, and conclusion about a cause-and-effect relationship between them (post-war analyses of Ajdukiewicz).

The difference in perspectives from which the two groups approached the issue may seem surprising due to the fact that the results of analyses performed by the second group (i.e. conclusion about the cause-and-effect relationship between perceptions and judgments) seem to deprive all considerations of the first group for taking place of cause–effect relation is possible between two separate beings, which can neither be a kind nor a component of one another. However, in my opinion, the above riddle can be solved by taking into account Twardowski's theory of acts and products. From the perspective of this theory, it can be seen that the first group dealt with judgments understood as acts (i.e. psychological acts), and the second with judgments understood as products (i.e. logical judgments and their linguistic expressions). In this respect, it can be recognised that the School has developed a solution to the question of the relationship between perceptions and judgments at once in two fields—of psychology and logic—which may certainly be regarded as an asset.

References

Adlerówna (Adler), M. (1935). Dwa rodzaje spostrzeżeń i ich struktura. *Kwartalnik Psychologiczny, 7*, 1–24.

Ajdukiewicz, K. (1931). O obiektywności poznania zmysłowego. *Ruch Filozoficzny, XII*, 212b–214a.

Ajdukiewicz, K. (1948). *Propedeutyka filozofii dla liceów ogólnokształcących*. Wrocław and Warszawa: Książnica-Atlas.

Ajdukiewicz, K. (1965). *Logika pragmatyczna*. Warszawa: PWN. English translation: Ajdukiewicz, K. (1974). *Pragmatic Logic*. Dordrecht and Boston: Reidel.

Baley, S. (1922). *Нарисисихольоґії*, Львів-Київ: Новішляхи. Reprint: Baley, S. (2002). Нарисписихольоґії. In С. Балей & М. Верников (Eds.),

Зібранняпраць у п'ятитомах. Т. 1 (pp. 271–353). Львів-Одеса: ІФЛІСАФС 'Cogito'.

Baley, S. (1935). *Zarys psychologii w związku z rozwojem psychiki dziecka*. Lwów and Warszawa: Książnica-Atlas.

Benjamin, A. (2007). Perception, judgment and individuation: Towards a metaphysics of particularity. *International Journal of Philosophy, 15*(4), 481–500.

Brentano, F. (2009). *Psychology from an empirical standpoint*. London and New York: Routledge.

Czeżowski, T. (1933). Spostrzeżenia i przypomnienia. *Kwartalnik Psychologiczny, 4*, 237–244.

Mehlberg, H. (1931). Zagadnienie obiektywności i adekwatności spostrzeżeń a interpretacja zdań spostrzegawczych. *Ruch Filozoficzny, XII*, 206a.

Olech, A. (1999). O kłopotach związanych z tetycznością. In W. Tyburski & R. Wiśniewski (Eds.), *Polska filozofia analityczna. W kręgu Szkoły Lwowsko-Warszawskiej* (pp. 207–216). Toruń: Wydawnictwo UMK.

Olech, A. (2000). W sprawie *poczucia realności świata*. In J. Hartman (Ed.), *Filozofia i logika. W stronę Jana Woleńskiego* (pp. 334–360). Kraków: Wydawnictwo „Aureus".

Oleksiuk, S. (1932). *O tzw. sądzie spostrzeżeniowym. Studium z psychologii poznania*. Lwów: typescript. Reprint (partial): Oleksiuk, S. (2015). O przyrodzie i przedmiocie tzw. sądu spostrzeżeniowego. *Studia z Historii Filozofii, 4, 6*, 53–74.

Paczkowska-Łagowska, E. (1976). Przedmowa tłumacza. In K. Twardowski (Ed.), Idea i percepcja. Z badań epistemologicznych nad Kartezjuszem. *Archiwum Historii Filozofii i Myśli Społecznej, 22*: 319.

Schetz, A. (2013). Uzasadnienie zdań spostrzeżeniowych: Ajdukiewicz, McDowell, Davidson. *Przegląd Filozoficzny – Nowa Seria, 22, 4*(88), 445–458.

Twardowski, K. (1892). *Idee und Perception. Eineerkenntnis-theoretische Untersuchungaus Descartes.* Wien: Carl Konegen. Polish translation: Twardowski, K. (1976). Idea i percepcja. Z badań epistemologicznych nad Kartezjuszem. *Archiwum Historii Filozofii i Myśli Społecznej, 22*, 321–344.

Twardowski, K. (1894). *Zur Lehrevom Inhalt und Gegenstand der Vorstellungen: einepsychologische Untersuchung*. Wien: Alfred Hölder. Polish translation: Twardowski, K. (1965), O treści i przedmiocie przedstawień. In K. Twardowski (Ed.), *Wybrane pisma filozoficzne* (pp. 3–91). Warszawa: PWN.

Twardowski, K. (1898). *Wyobrażenia i pojęcia*. Lwów: H. Altenberg. Reprint: Twardowski, K. (1965). Wyobrażenia i pojęcia. In Twardowski, K. (Ed.), *Wybrane pisma filozoficzne* (pp. 114–197). Warszawa: PWN.

Twardowski, K. (1923). O *naukach apriorycznych czyli racjonalnych (dedukcyjnych)* i *naukach aposteriorycznych czyli* empirycznych (indukcyjnych). In K. Ajdukiewicz (Ed.), *Główne kierunki filozofii w wyjątkach z dzieł ich klasycznych przedstawicieli* (pp. 180–190). Lwów: K.S. Jakubowski.

Witwicki, W. (1925). *Psychologia do użytku słuchaczów wyższych zakładów naukowych* (Vol. 1). Lwów, Warszawa, and Kraków: Wydawnictwo Zakładu Narodowego im. Ossolińskich.

8

On the Lvov School and Methods of Psychological Cognition

Teresa Rzepa

1 Introduction

It's not difficult to demonstrate that Polish psychology derives mainly from two trends which dominated Europe at the end of the nineteenth century. The first trend, related to the academic activity of Wilhelm Wundt (1832–1920), led to establishing experimental and physiological psychology, manifested by 'trendy' psychological laboratories, departments and studies, initiated on a massive scale. The second trend, focused on searching for the 'royal' road to awareness, should be associated with the descriptive psychology of Franz Brentano (1838–1917).

These trends were encountered by key representatives of the first generation of Lvov psychologists—Władysław Witwicki (1878–1948), Stefan Baley (1885–1952), Stefan Błachowski (1889–1962),

T. Rzepa (✉)
Department of Psychology in Poznań,
University of Social Science and Humanities, Warsaw, Poland
e-mail: trzepa@swps.edu.pl; trz@data.pl

A. Drabarek et al. (eds.), *Interdisciplinary Investigations into the Lvov-Warsaw School*, History of Analytic Philosophy,
https://doi.org/10.1007/978-3-030-24486-6_8

Mieczysław Kreutz (1893–1971). They chose the latter trend, consistently following the master and founder of the Lvov-Warsaw School, Kazimierz Twardowski (1866–1938).

After 1918, the above mentioned disciples of Twardowski became Chairs of all faculties of psychology at Polish universities (apart from the Jagiellonian University), instilling 'Lvov' ideas in subsequent generations of psychologists.

In 1895, the 29-year-old Twardowski became Chair of the Faculty of Philosophy at the University of Lvov. It was not difficult for him to win 'the hearts and minds' of students who were almost his age. He was able to attract them with his dynamic nature, readiness for café meetings and never-ending discussions, respect-commanding personality and vast knowledge—not only about philosophy. He would talk eagerly of psychology as a new science, he knew how to reach people's psyche, he was fascinated by ethics and methodology, applied 'explicitly clear' principles of reasoning and expressing thoughts.

He rejected 'national' philosophies, all 'messianisms and mysticisms', treating them as private visions of the world, worthless for academic analyses. According to him, science meant precise definitions and analyses, reliable interpretations, emotionless consideration, rational approach, adequately selected research methods, logical and clear products. All that because only well-organised thinkers are able to express themselves clearly, whereas vague thoughts do not deserve to be uttered, or to be paid heed to by any listeners or readers (Rzepa 1998).

Due to his valuable, though often underestimated proposals, Twardowski and his disciples left an indelible mark on Polish psychology. The characteristic trait of the Lvov-Warsaw School, also in its psychological current, was the method of cognition which will be discussed in this article. It was concerned with the idea and applicability of introspection, the method of reconstructing psychological facts and observation of life, psychological interpretation, opposition to the test mania, the model of relationship between the psychologist and the examined person and a linguistically impeccable yet vivid style of academic writing.

Twardowski's School may also boast to be the prime movers in the area of psychology, not only in Poland. For example, the animated discussions held at the time with regard to understanding the object of psychology as 'mental life' were crystallised in Twardowski's theory of actions and products (Twardowski 1912/1965), justifying the legitimacy of psychological interpretation which is typical to psychoanalysis and humanistic psychology.

Witwicki (1900, 1907) created the theory of kratism that preceded Alfred Adler's (1870–1937) theory of will to power (Markinówna 1935). Kratism explains the genesis of humour, the effect of first impression and development of social feelings, indicating their functions which regulate human relations, and enabling the reception of cultural products. It also allows for typologizing people depending on their vital powers and attitude to others. It is feelings that constitute a bridge linking the world of science with the world of art, culture, politics and religion, as they have a modifying effect on perception, memory, thinking, learning or acts of will.

In 1909, Witwicki was the first person ever to draw up a psychobiography of Socrates, which preceded the psychobiographic presentation of Leonardo da Vinci by Sigmund Freud (1910/1975). It was also Witwicki (1939/1980) who described the idea and genesis of supposition, that is, 'quasi-beliefs', when he was researching the faith of 'enlightened' persons, trying to answer the question about how it is possible that educated people having a rational system of knowledge about the world simultaneously believe in irrational religious dogmas. The research led him to discover the psychological rule of conflict, which preceded the theory of cognitive dissonance published in 1957 by Leon Festinger (1919–1989).

Błachowski (1928) was one of the first European psychologists to propagate applied psychology, using psychoanalysis to explain 'mental epidemics'. In cooperation with Stefan Borowiecki (1881–1937), he performed experiments in which people experienced states of ecstasy and religious hallucinations, which indicated the psychological nature of the phenomena and their pathological origin. Both researchers stated that it was a low intelligence level, susceptibility to hysteric behaviours and external influence as well as underdeveloped criticism that predisposed people to ecstatic experiences and religious visions. Moreover, they found that the epidemics of ecstasies should be attributed to increased suggestiveness which triggered a mechanism of regression to an infantile state and magical thinking.

Baley discovered the phenomenon of syncretic perception found in young children. Moreover, he contributed to the development of psychology as the author of the first textbooks on educational psychology connected with social psychology, indicating the previously unnoticed sources of education (Baley 1931, 1938/1958). He was the first author in Poland to describe the development of concepts, ethical attitudes and artistic creativity in children and the acquisition of social behavioural patterns. Moreover, he introduced the concept of 'personality' as an important term in Polish psychology.

And last but not least, other contributors to major academic achievements included Kreutz (1962), whose apt defence of the scientificity of introspection remains valid even today, and Witwicki (1925/1962, 1927/1963) *who authored the first* two-volume textbook on general psychology to be published in Poland.

2 Background

Since Twardowski took numerous and effective measures to promote psychology right from his arrival in Lvov until his death, he should also be recognised as the founder of a school of psychology. The Lvov School of Philosophy, in view of its historical and factual continuity owing to Twardowski's disciples who were employed at the University of Warsaw after 1918, is also named the Lvov-Warsaw School (Woleński 1985). However, the school of psychology, whose genesis and development are connected exclusively with Lvov, should be referred to as the Lvov School. The history of this academic formation can be divided into the following periods: (1) from the turn of 1898–1899 to 1901; (2) from 1902 to 1919; and (3) from 1920 to 1939 (Rzepa 1998, 2002, 2016; Rzepa and Dobroczyński 2009).

In the academic year 1898–1899, Twardowski delivered lectures in psychology where he discussed visual delusions, using tables to illustrate the discussed phenomena. Soon afterwards, he claimed that their significance was in fact much greater, referring to his presentation as a psychological experiment, and officially announced (Twardowski 1904) the establishment of the first psychology laboratory in Poland (Rzepa and Dobroczyński 2009). Even though this statement was far-fetched, from then on psychology stayed at the Lvov University for good.

Motivated by the growing interest in a new field of science, Twardowski made psychology classes a permanent part of the philosophy curriculum. At that time, he published Polish versions of texts on the objectives and methods of psychology and its relation to other disciplines, especially philosophy and physiology (Twardowski 1897/1965, 1898/1965).

He introduced Polish terms (along with their definitions) to describe key psychological concepts, such as 'sensation', 'mental image', 'perception', 'representation', 'mental disposition'. In 1901, he was given permission by Galician authorities to use several rooms to run psychological experiments (Twardowski 1913/1965; Rzepa 1998; Stachowski 2000).

This period ended on April 14, 1901, when Witwicki (1900) defended the first doctoral dissertation in psychology presented in Lvov, which included the key assumptions of the kratism theory.

A number of facts suggest that the prime years of Lvov psychology were the period from 1902 to 1919. Twardowski had won the arduous struggle to obtain regular subsidies for the psychology laboratory, which as from 1907 became an integral part of the University. In the years 1901–1902, with a grant arranged by Twardowski, Witwicki attended lectures in descriptive psychology in Vienna; he also studied in Lipsk, where he worked at the famous laboratory founded by Wundt in 1879.

In 1904, Witwicki (as the first from among Twardowski's disciples) became an independent researcher and began regular cooperation with the University as a private associate professor. At that time, he developed a theory of kratism, treating ambition as the 'will to power' and a major motivator of human actions (Witwicki 1907). He distinguished between two states connected with the experience of a sense of power, an 'elevating' and a 'diminishing' one, which dominate interpersonal relationships. Kratic aspirations manifest themselves in continuously striving towards 'a sense of power' and avoiding 'a sense of powerlessness'. A drive for power may be satisfied by diminishing oneself or others, or by elevating oneself or others. The will to power principle governs human behaviour and depends on the course of individual development. A person who develops properly will become independent, equal or superior in relation to their social environment. Otherwise, they will become dependent and inferior to/weaker than others.

In 1911, Baley defended his doctoral dissertation under Twardowski's supervision. Later on (1912–1914), he availed himself—also owing to arrangements made by his master—of scholar grants in Berlin (under Carl Stumpf) and in Paris, and after returning to Lvov he then completed medical studies, obtaining a Ph.D. in 1923. Thanks to his education, he found a way to combine psychology and medicine and to practise psychology.

In 1912, Twardowski proposed that actions and products should be divided into physical and mental ones, and into permanent and impermanent ones, where the permanent trait was attributed to physical actions and products. Among them, he distinguished psychophysical actions and products of specific origin, seeing that a psychophysical action occurs when a permanent physical action is accompanied by an impermanent mental action that influences the course of the former, and consequently the resulting psychophysical product, e.g. a letter, a speech, a drawing, a psychophysical document, a painting, a piece of music, a diary…

Psychophysical products may be permanent or impermanent, but usually they can be observed and interpreted, thus evoking some specified mental products in the observers and interpreters which are analogous to those experienced by their authors in the course of their creation.

In 1913, Stefan Błachowski obtained his Ph.D. degree under the supervision of Georg Elias Müller (1850–1934). After returning to Lvov and taking the position of assistant to Twardowski, Błachowski obtained his post-doctoral degree in 1917, defending his dissertation entitled 'Attitudes and Observations. A psychological Study'. Apart from his academic work, he involved himself in practical issues, such as improving memory, visual thinking and learning, or measuring mathematical ability and other skills.

In that period, Twardowski continued enriching the language of Polish psychology, mainly in relation to translation of psychological texts. He was supported by Witwicki, Błachowski and Kreutz. It was then that the works of famous psychologists such as William James, Wilhelm Wundt, Gustav Th. Fechner, Alois Höfler, August H. Forel, Theodule A. Ribot were translated into Polish. In addition, the propagation of current psychological academic events was facilitated by the *Ruch Filozoficzny* ('Philosophic Movement') quarterly founded by Twardowski in 1911.

The peak period in the development of the Lvov School ended in 1919, when thanks to Twardowski's efforts, his disciples became Chairs of Psychology Faculties at four universities. On July 16, 1919, Błachowski was appointed Professor of Psychology at the newly founded Poznań University. Witwicki received his nomination on April 1, 1919 and became Chair of the Faculty of Psychology at the University of Warsaw. Following long discussions on setting up a second Faculty of Psychology at the University of Warsaw, and after founding the Faculty of Educational Psychology, Baley was appointed Associate Professor (October 16, 1928). Moreover, at Stefan Batory University in Vilnius, lectures in psychology were delivered by Błachowski, and then his doctoral student, Rev. Mieczysław Dybowski (1885–1975) who was also employed at the Poznań University; then (in 1935) Bohdan Zawadzki (1902–1966), a disciple and doctoral student of Witwicki, became Head of the Psychology Department in Vilnius (Rzepa 2002).

In the years 1920–1939, the fortunes of the Lvov School of Psychology became increasingly complicated due to the dispersion of Twardowski's disciples. Still, this also provided an opportunity to replicate the Lvov model and to establish their own schools of psychology. However, it did not turn out this way. In its final years, the Lvov School functioned as though beyond the universities, due both to Twardowski's work and to the transfer of ideas typical for the School as regards the object of psychology, its tasks, methods and applications. This situation was enhanced by frequent meetings, the exchange of letters and mutual support.

In 1920, Twardowski's efforts resulted in establishing the Institute of Psychology. He chaired the Institute until 1928, while urgently looking for a successor, and he encouraged Kreutz to speed up work on his post-doctoral dissertation ('Variability of Tests Results'). Kreutz defended his post-doctoral thesis on December 7, and delivered his post-doctoral lecture on December 10, 1927. On June 30, 1928 he was appointed Associate Professor and was entrusted the Psychology Institute and Laboratory (Twardowski 1997: 2).

In that period, Twardowski and Kreutz educated promising psychologists who—alas—were killed during World War II. They included Walter Auerbach (?–1944), Eugenia Blaustein (1905–1944) and Leopold Blaustein (1905–1944), who made a major contribution to the psychology of perception, analysing the 'representation' category in relation to attitudes and judgement. His findings became the basis for classifying aesthetic experiences and reflections on the perception of radio dramas and films. Moreover, Andrzej Lewicki (1910–1972), who developed mainly under Kreutz's guidance, was later to organise the teaching of psychology at Nicolaus Copernicus University, and establish the first Chair of Clinical Psychology in Poland (at Adam Mickiewicz University). Another of Kreutz's students was Tadeusz Tomaszewski, who followed the curriculum of organisational and didactic activities introduced by Twardowski. In fact, Tomaszewski was the only actual founder of a school of psychology, gathering students around his own theory of actions (Rzepa and Stachowski 2011). However, from today's perspective, that chapter is now closed.

For obvious reasons, the final period ended in 1939 with the outbreak of World War II. It is now time for a bitter reflection: even though Twardowski's disciples had the good fortune to meet an exemplary master figure—a scholar, an educator, an organiser—who inspired and attracted young people, no one was able to replicate his success. For this reason, the Lvov School of Psychology might be called 'a school of wasted opportunities' (Rzepa 1998, 2002).

3 Methods of Psychological Cognition

One of the strengths of the Lvov School of Psychology was its method of research focused on increasingly more effective exploration of mental life as the basic object of psychology. According to the state of knowledge at that time, mental (spiritual, internal) life comprised of facts (phenomena) and mental dispositions (e.g. sensitivity, memory, imagination, intelligence, will, character). The direct and obvious source of knowledge about mental life was inner experience—conscious sensations and impressions—which are always 'someone's', always felt by 'someone'. Accordingly, mental life is subjective and non-sensory, and thus impossible to be controlled or observed by others. Introspection is the only way for an adequately trained psychologist to gain an insight and acquire knowledge about the quality and dynamics of phenomena, dispositions, processes and changes taking place in the area of mental life (Witwicki 1902).

In one of the first texts on the object and methods of psychology, based on the findings of his master, Franz Brentano, Twardowski (1897/1965) distinguished between psychological and physiological methods, that is, internal perception (experience) and external perception (observation). He demonstrated that introspection should not be equated with self-observation, as: (1) mental phenomena occur too fast and last too short for anybody to be able to observe them in a reliable manner and (2) conditions required for scientific observations to be correct are impossible to be satisfied when the observer is simultaneously the object of observation.

Consequently, Twardowski found that while introspection as inner experience is admittedly a source of direct knowledge about the psyche, it is not free from certain drawbacks. There are aspects of introspection that cannot be overcome: (1) it is limited to the mental life of a single person (unsurmountable subjectivity) and (2) mental facts cannot be experienced concurrently with their reliable and systematic observation. As for the latter drawback, Twardowski (1913/1965) suggested that it could be mitigated by way of an experimental method, employed by German psychophysiologists since the times of Fechner (1801–1887), and consisting in the registration of somatic symptoms accompanying mental phenomena (Wundt 1908).

In order to eliminate the former drawback of introspection, Twardowski came up with an entirely original idea of his own. He put forward a theory of actions and products (Twardowski 1912/1965), making it possible to overcome the insurmountable subjectivity of introspection by interpreting facts of somebody else's mental life. He claimed they could be reconstructed from psychophysical products, that is, retained mental products (thoughts, feelings, mental images, desires, fantasies and the like), which were formed in the minds of their creators.

Such products could be obtained by applying an experiment involving introspection which was successfully employed by the founder of the Würzburg School, Oswald Külpe (1862–1915). The method consisted in repeated, systematic and purposeful elicitation of certain mental phenomena (using memory and imagination), in order to treat them as objects of observation. The person who re-experienced the phenomena then provided a detailed oral or written account describing the cognitive processes involved in the direct and internal experience. The introspection experiment, whose application was strongly objected to by Wundt, was based on a *post factum* strategy, that is, the reconstruction of already experienced mental phenomena. We could say that Külpe's proposals sanctioned the findings resulting from Twardowski's theory of actions and products. After all, during the reconstruction of one's own or someone else's mental facts, a psychologist should treat them as psychophysical products.

The psychologist should thus re-invoke and observe the mental actions that bring about the products, and decipher the fleeting mental products, existing in the mind of their author at the moment when the currently reconstructed psychophysical products were being made. To make sure the process proceeds correctly and is not distorted by any difficulties connected with empathising with someone, one must be skilful in applying analogy and introspection (Twardowski 1913/1965).

Moreover, mental products often become endowed with the tangible form of psychophysical products when they are transformed into psychological documents. These include, *inter alia*, diaries, autobiographies, letters, opinions and results of psychological tests, literary works, paintings, music pieces in which mental products are captured regardless of the passage of time and place of storage. Psychological interpretation of somebody else's psychophysical products is an extremely difficult task, since it requires appropriate resources of knowledge and skilful use not only of analogy and introspection, but also of psychological intuition. This ability is only displayed by good (according to Witwicki's standards) psychologists who are able to reach the depths of a person's psyche even when their psychological knowledge fails.

Only good psychologists are able to analyse psychological documents and expand their knowledge about mental life with understanding resulting from their learning, intuition, competence in interpretation and analogy-based experience of other people's mental products. They are entitled to do that, as this way of applying the method minimises objections regarding the insurmountable subjectivity of introspection. This kind of psychological knowledge, arrived at through the so-called life-observation method, was successfully applied by 'the good psychologist' Witwicki (1927/1965, 1939/1980), who carefully interpreted the observed and/or documented psychophysical products (Rzepa 2002).

The theory of actions and products provides the scientific foundations for seeing the act of interpretation as an important form of psychological activity. Firstly, due to the fact that a similar or identical meaning of a psychophysical product can be arrived at by any number of interpreters, it is possible to see that mental actions performed by different persons at different times and in different places lead to the creation of a similar mental product. This means that the mental product becomes in a way 'independent' from the subject's action from which it sprang, and thus continues to exist in the psychophysical product, 'waiting' to be discovered or disclosed once again. Secondly, since it is possible to ignore the strictly subjective characteristics of mental products, it becomes permissible to disregard the influence of interpreters on the process of interpretation itself. Therefore, interpretation appears to be a *quasi*-objective procedure (Rzepa 2002). Kazimierz Ajdukiewicz (1890–1963) put it this way in the unique style of the Lvov-Warsaw School:

> Human thoughts are ephemeral, they are born and die within a short time, as soon as they vanish from our consciousness. However, there is something in them that may last for centuries in a potential form: the contents of thoughts. If a thought is expressed in words, then the content of the thought, as the linguistic meaning of the words, remains connected with them, and words may last for years on end. […] Therefore, the contents of thoughts […] enjoy an existence that is independent from the fortunes of those who produced them in their thoughts. (Ajdukiewicz 1985: 307–308)

Another methodological achievement of the Lvov School of Psychology, resulting from its position on the applicability of introspection, is the specific attitude to the model of research prevailing in psychology at the time. Although Twardowski and his disciples came in touch with psychophysiology and psychophysics during their work at German laboratories, even though they deemed experiment to be a psychological method as well, useful especially for the purposes of authenticating the results of introspection, and despite the fact they constructed and used an apparatus for measuring physiological processes which accompany cognitive processes, they rejected the practice of psychology in the form implemented by Wundt.

The Lvov laboratory functioned on different principles, the most significant of which were the ones which required mastering the skill of psychological description (Brentano 1874/1999) and interpretation. Another important principle which distinguished the Lvov laboratory was its rejection of the trendy model of psychological investigation, which as from the 1900s was unquestioningly introduced and become vastly popular in psychology: the virtually unlimited production of tests on almost 'any occasion'. Due to the explicit stance taken by Witwicki (1928, 1929), who pointed out their unreliability and inaccuracy, as well as the fact that they ignored cultural differences, Polish psychologists took a conservative and cautious approach to the test mania.

This psychological junk first entered Polish schools, where tests were used to measure the students' intelligence by asking them to list at least ten American presidents, or answer questions like: 'Do figs grow in trees or bushes?', or 'Is a dromedary an interesting animal?' These intelligence tests also included questions that were simply disrespectful to the students' intelligence: pupils were asked to indicate whether the following statements were true or false: 'Flowers bloom'; 'Apples are edible'; 'Newspapers are printed in churches'; 'Theft is a right thing to do'; 'A horse has five legs'; 'Frogs have wings'. The authors of those thoughtlessly translated tests did not seem to notice the ridiculous paradoxes, in reaction to which Polish students showed their sense of irony. For example, faced with a gap-filling exercise such as: 'In this school students have their (*classes*) until 5:00 p.m.', a number of students wrote: 'In this school students have their *lunch break* until 5:00 p.m.', Of course, such answer was counted as incorrect (Witwicki 1928).

Thanks to the critical attitude of Lvov psychologists to the trendy, though diagnostically inadequate, methods of psychological cognition, Poland was saved from the 'test flood'. On the other hand, the commotion around the test mania had some positive effects, as it changed the attitude of academic psychologists to psychological practice and initiated actions aimed at establishing the profession of psychologist. In this context, it was necessary to specify the model of communication between psychologists and their patients. Although neither Twardowski nor his disciples practised psychotherapy, he was the first in Poland to create a model of communication between the psychologist and the examined person (Twardowski 1912/1965; Rzepa 2002). The model involves: (1) a psychologist who interprets the products and reconstructs the mental life of their patient; (2) a mental product (manifestation, symptom) or a permanent, interpretable psychophysical product (psychological document); and (3) the creator of products. The indispensable elements of this model are the products and the psychologist who interprets them, whereas the creator of the product is dispensable. The proposed model assigned to the psychologist the role of a non-interfering, objective observer and interpreter, whose main task was to obtain and interpret psychological documents. Obviously, the model was far from representing the dynamics of the psychotherapeutic process; nevertheless, attention was drawn to the specific nature of the relationship between the psychologist and the examined person (patient).

Another methodological achievement of the Lvov School of Psychology was the analytical method in its descriptive variant, applied mainly to define psychological concepts in accordance with the model developed by Brentano, saying that a phenomenon must first be clearly, explicitly and reliably described before it can be explained.

> A thoughtful, accurate analysis of even a single fact is enough to formulate a general statement, and then we only need to verify what we have discovered based on an identical or even very similar experience. This is the descriptive analytical method of psychology that was developed by Twardowski himself, and established in psychology through his [...] disciples. (Słoniewska 1959: 21)

Twardowski's uncompromising insistence on the clarity of language used in science is well-illustrated by the famous project of the Lvov-Warsaw School, named the 'ABC of Proper Thinking' (Ajdukiewicz 1959). It is a set of rules designed to define and analyse the meaning of concepts, ensure consistency in the use of scientific terms and provide logical substantiation of propositions in a rational and unemotional way. 'What manifests the objective nature of scientific research is that it never follows orders from external factors, never serves any subsidiary causes, recognises experience and reasoning as its only masters, and serves only one task: to pursue properly substantiated true statements [...]' (Twardowski 1932/1992: 462).

The style of scientific writing derived from the rules of 'ABC of Proper Thinking' characterises the works of most academics of Lvov origin. The impact and lasting effect of Twardowski's teaching is illustrated, for instance, by a letter Witwicki wrote to his son (June 23, 1937):

> One of the purposes served by science is to set boundaries to the meaning of words, which are non-negotiable within its realm. This is done by means of definitions. [...] Of course, we are always allowed to go beyond the boundaries defined by scientific definitions or linguistic customs if we want to deliberately cause a misunderstanding, for example for the purposes of jokes, mischief [...] or metaphors. [...] This is allowed, as long as we remember that we are always running the risk of misunderstanding unless we explicitly agree beforehand that we are not going to stick to set terms. Well-defined terms enable clear thinking.

In 1920, Twardowski (1919–1920) took a stand on the clarity (or vagueness) of thoughts expressed with words, and reprimanded all long-winded authors who displayed a woolly style of writing, used jargon and vague concepts, muddled up illogical threads, etc. Moreover, he harshly stated that only those who could express themselves clearly could think clearly. In his opinion, vagueness and elaborateness of the language did not manifest depth of thought. Just the opposite: such traits were evidence of a thoughtless void concealed under a delusion of depth. In addition, Twardowski argued that it was not worth anybody's while to decipher the thoughts of authors who were unable to express themselves clearly and concealed platitudinous contents behind a smokescreen of elaborately ornate disquisitions. Similarly, he expressed his contempt about symbolomania, that is, the tendency for excessive use of symbols, and pragmatophobia—intellectual aversion to using clear and understandable language in communication (Twardowski 1921).

Psychologists descended from Twardowski's School received methodological guidance on the nature and applicability of introspection, methods of reconstructing mental facts and life observations, the value of tests, the manner of providing reliable descriptions and apt interpretation of people's products, as well as the specific nature of the relationship between the psychologist and the examined person. It was due to the methodological soundness of the School that its psychologists adopted an inquisitive approach to the subject matter and tools of psychological cognition; moreover, they were taught to take responsibility for its effects, which contributed to their better understanding of mental life. The School failed, however, to teach them how to boldly disseminate the effects of their works beyond Poland.

4 In Lieu of a Conclusion

The words Twardowski spoke on November 21, 1932 at the lecture hall of Jan Kazimierz University, where he was awarded honorary doctorate, sound surprisingly relevant. They were uttered by a person who had made an immense contribution to the development of Polish science, who considered universities to be ideologically independent sanctuaries of science and whose life and work illustrated the meaning of noble work and invincible faith in the power of science which strives to discover truth by employing methods of scientific inquiry.

> How much of the hatred which divides humankind into bitter adversaries would cease if disputes from which it springs were examined using scientific methods! It would then be possible either to resolve the disputes objectively, or else we would find that the conflicting opinions are equally groundless, and therefore none of the parties has the right to demand that their ideas be embraced. Objective truth, and even the mere act of pursuing it with all due diligence, helps relieve the disputes and altercations between adversaries. It removes that which inspires people against one another, and offers that which reconciles them, thus making them look favourably upon each other. Thus, the pursuit of objective truth becomes endowed with an ethical dimension, and may increasingly become a true blessing for humankind. (Twardowski 1932/1992: 463)

References

Ajdukiewicz, K. (1959). Pozanaukowa działalność Kazimierza Twardowskiego. *Ruch Filozoficzny, 19*, 29–35.

Ajdukiewicz, K. (1985). *Logika pragmatyczna*. Warszawa: Państwowe Wydawnictwo Naukowe.

Baley, S. (1931). *Psychologia wieku dojrzewania*. Lwów and Warszawa: Książnica-Atlas.

Baley, S. (1938/1958). *Psychologia wychowawcza w zarysie*. Warszawa: Państwowe Wydawnictwo Naukowe.

Błachowski, S., & Borowiecki, S. (1928). Epidemia psychiczna w Słupi pod Środą. *Rocznik Psychiatryczny, 7*, 176–197.

Brentano, F. (1874/1999). *Psychologia z empirycznego punktu widzenia*. Warszawa: Państwowe Wydawnictwo Naukowe.

Freud, S. (1910/1975). *Leonarda da Vinci wspomnienie z dzieciństwa*. Warszawa: Państwowe Wydawnictwo Naukowe.

Kreutz, M. (1962). *Metody współczesnej psychologii*. Warszawa: Państwowe Zakłady Wydawnictw Szkolnych.

Markinówna, E. (1935). Psychologia dążenia do mocy. Zestawienie poglądów Witwickiego i Adlera. *Kwartalnik Psychologiczny, 7*, 329–340.

Rzepa, T. (1998). *Życie psychiczne i drogi do niego. Psychologiczna Szkoła Lwowska*. Szczecin: Wydawnictwo Naukowe Uniwersytetu Szczecińskiego.

Rzepa, T. (2002). *O interpretowaniu psychologicznym. W kręgu Szkoły Lwowsko-Warszawskiej*. Warszawa: Polskie Towarzystwo Semiotyczne.

Rzepa, T. (2016). Kazimierz Twardowski i psychologiczna Szkoła Lwowska. In A. Brożek & A. Chybińska (Eds.), *Fenomen Szkoły Lwowsko-Warszawskiej* (pp. 87–110). Lublin: Wydawnictwo Academicon.

Rzepa, T., & Dobroczyński, B. (2009). *Historia polskiej myśli psychologicznej. Gałązki z drzewa Psyche*. Warszawa: Wydawnictwo Naukowe PWN.

Rzepa, T., & Stachowski, R. (2011). O tym, kto jest mistrzem dla polskich psychologów i jaki ów mistrz jest. W: T. Rzepa, C.W. Domański (red.), *Na drogach i bezdrożach historii psychologii* (pp. 251–264). Lublin: Wydawnictwo UMCS.

Słoniewska, H. (1959). Kazimierz Twardowski w psychologii polskiej. *Ruch Filozoficzny, 19*, 17–24.

Stachowski, R. (2000). *Historia współczesnej myśli psychologicznej. Od Wundta do czasów najnowszych*. Warszawa: Scholar.

Twardowski, K. (1897/1965). Psychologia wobec fizjologii i filozofii. In *Wybrane pisma filozoficzne*. Warszawa: Państwowe Wydawnictwo Naukowe.

Twardowski, K. (1898/1965). Wyobrażenia i pojęcia. In *Wybrane pisma filozoficzne*. Warszawa: Państwowe Wydawnictwo Naukowe.

Twardowski, K. (1904). List oficjalny do redakcji „Przeglądu Filozoficznego". *Przegląd Filozoficzny, 7*, 104.

Twardowski, K. (1912/1965). O czynnościach i wytworach. In *Wybrane pisma filozoficzne*. Warszawa: Państwowe Wydawnictwo Naukowe.

Twardowski, K. (1913/1965). O psychologii, jej przedmiocie, zadaniach, metodzie, stosunku do innych nauk i o jej rozwoju. In *Wybrane pisma filozoficzne*. Warszawa: Państwowe Wydawnictwo Naukowe.

Twardowski, K. (1919–1920/1965). O jasnym i niejasnym stylu filozoficznym. In *Wybrane pisma filozoficzne*. Warszawa: Państwowe Wydawnictwo Naukowe.

Twardowski, K. (1921). Symbolomania i pragmatofobia. *Ruch Filozoficzny, 6*, 1–10.

Twardowski, K. (1932/1992). O dostojeństwie uniwersytetu. In K. Twardowski (Ed.), *Wybór pism psychologicznych i pedagogicznych* (pp. 461–471). Warszawa: Wydawnictwa Szkolne i Pedagogiczne.

Twardowski, K. (1997). *Dzienniki* (Vols. 1–2) (Oprac. R. Jadczak). Toruń: Wydawnictwo Adam Marszałek.

Witwicki, W. (1900). Analiza psychologiczna ambicji. *Przegląd Filozoficzny, 3,* 26–49.

Witwicki, W. (1902). Rzut oka na kierunek W. Wundta w dziejach filozofii. *Tygodnik Słowa Polskiego, 9,* 10.

Witwicki, W. (1907). Z psychologii stosunków osobistych. *Przegląd Filozoficzny, 10,* 531–537.

Witwicki, W. (1909). Wstęp. Objaśnienia. In Platon (Ed.), *Uczta.* Lwów: Książnica-Atlas.

Witwicki, W. (1925/1962). *Psychologia* (Vol. 1). Lwów: Wydawnictwo Zakładu Narodowego im. Ossolińskich.

Witwicki, W. (1927/1963). *Psychologia* (Vol. 2). Lwów: Wydawnictwo Zakładu Narodowego im. Ossolińskich.

Witwicki, W. (1928). O narodowych testach amerykańskich do badania inteligencji. *Psychotechnika, 2*(7), 23–32.

Witwicki, W. (1929). Odpowiedź p. M. Kaczyńskiej. *Psychotechnika, 3,* 33.

Witwicki, W. (1939/1980). *Wiara oświeconych.* Warszawa: Państwowe Wydawnictwo „Iskry".

Woleński, J. (1985). *Filozoficzna szkoła lwowsko-warszawska.* Warszawa: Państwowe Wydawnictwo Naukowe.

Wundt, W. (1908). *Grundzüge der physiologischen Psychologie* (Vol. 3). Leipzig: Engelmann.

9

The Interdisciplinary Nature of Władysław Witwicki's Psychological Investigations

Amadeusz Citlak

1 Introduction

Władysław Witwicki is one of the closest and best-known students of Kazimierz Twardowski, founder of the Lvov-Warsaw School. His psychological achievements are still little known in international literature on psychology, however. This is surprising, given that his research was very original and could very well be included in the mainstream of the most important theoretical problems of psychology in the first half of the twentieth century. That this did not happen was mainly due to political reasons (restrictions imposed on the School and its discontinuation after the First and Second World War) and ideological

Scientific paper prepared as part of the MINIATURE 2 (National Science Centre, Poland, 2018/02/X/HS6/00278).

A. Citlak (✉)
Institute of Psychology, Polish Academy of Sciences, Warsaw, Poland
e-mail: acitlak@wp.pl

A. Drabarek et al. (eds.), *Interdisciplinary Investigations into the Lvov-Warsaw School*, History of Analytic Philosophy,
https://doi.org/10.1007/978-3-030-24486-6_9

trends (after 1918, Polish psychologists published mainly in the Polish language).

In this paper, I will draw attention to Witwicki's most original achievements and their similarity/parallels to the then contemporary achievements in the field of psychology. They involved a number of areas, in both the theoretical and the empirical tradition. Firstly, Alfred Adler's individual psychology and his theory of striving for power. Witwicki's original theory of cratism which is similar in its assumptions to the power theory, preceded Adler's work by several years. Even so, it remains virtually unknown in world literature. Secondly, the Dorpat School of the Psychology of Religion, whose achievements coincided in some respects with Witwicki's research into religious beliefs. Thirdly, Leon Festinger's theory of cognitive dissonance, and fourthly—drawing on the cratism theory—the psychobiographies of Socrates (already in 1909) and Jesus Christ (1958), which are some of the earliest non-psychoanalytical psychobiographies in the world.

Witwicki's psychological research was interdisciplinary and concerned with important problems of contemporary psychology. His investigations were also part of the historical (or cultural-historical) psychology developed by Wilhelm Wundt. One of the most important features of his analyses was the description and interpretation of empirical data, and, to a lesser extent, experiments and methods derived from natural sciences.

2 The Theory of Cratism and Alfred Adler's Individual Psychology

The most evidently parallel scientific tradition was the individual psychology of Alfred Adler and his theory of pursuit for the sense of power (or later 'perfection'). As is well known, Witwicki introduced elements of the cratism theory for the first time in his Ph.D. thesis entitled 'Psychologiczna analiza ambicji' ('A Psychological Analysis of Ambition') (Witwicki 1900), then during a lecture at the Congress of Physicians and Naturalists in Lvov (Witwicki 1907), and subsequently in the second volume of his Handbook of Psychology (Witwicki 1927).

His Ph.D. dissertation is considered to be a prototype of the cratism theory, as it contains its key theses, above all the concept of ambition, which later became something like the driving force behind cratic desires. Witwicki introduced the concept of cratism in 1907. He saw it realised in four different ways: by elevating oneself, by elevating others, by diminishing oneself and by diminishing others. Twenty years later, he performed a comprehensive analysis of social relations based on cratic desires, and developed an original typology of 'heteropathic' emotions, i.e., those experienced with respect to other people. The theory of cratism allowed Witwicki to explain and interpret various areas of human activity: the phenomenon of comic, religious and artistic activity (including the perception of art), and generally the dynamics of social relations. However, in some cases his application of the cratism theory went beyond the standards of reliable scientific work, and led the author to over-interpreting—to say the least—and contaminating his research material with preconceived assumptions.

Perhaps the best-known example is the cratic psychobiography of Jesus Christ, which I will mention further on in this paper. Irrespective of some weaknesses inherent in the theory itself, it is undoubtedly an original achievement of the Polish psychologist, and has a high explanatory and predictive value. What is most interesting, however, is that a strikingly similar theory was announced almost at the same time by a student of Freud's, Alfred Adler. He discussed the basis of *Individualpsychologie* in his works titled 'Studie über Minderwertigkeit von Organen' (Adler 1907), 'Über den nervösen Charakter' (Adler 1912) and 'Praxis und Theorie der Individualpsychologie' (Adler 1920). Its similarity to the theory of the Austrian psychologist was emphasised from the very beginning, for the first time in Polish literature in the work of Estera Markinówna (1935), and then in other studies and mentions of the cratism theory (Jadczak 1981; Nowicki 1982; Rzepa 1991). The analogies are so many that in my opinion both theories can be considered complementary.

Firstly, in both cases the basic motivation of human activity is the striving for power. Witwicki speaks about cratic desires, referring to ancient Greek literature uses the word 'kratos' (to be strong, powerful, take possession). Adler talks about striving for power (*Machtstreben*) or a sense of power (*Machtgefühl*). Power can be understood in the literal sense as strength and physical advantage, or as something acquired through competences, skills or resources (which usually applies to social relations). Although Adler places more emphasis on the sense of power resulting from competence, and even identifies it with the concept of perfection (*Vollendung, Vollkommenheit*), it ultimately remains in line with Witwicki's assumptions, in which social competence and striving for perfection play a similar role. For both psychologists, a sense of power can also result from a subjective view of things, and does not have to be grounded in facts.

Secondly, a sense of inferiority plays an essential role in both theories as a factor that may trigger the desire for a sense of power. However, Witwicki understood cratic desires as a biologically conditioned instinct, independent of social influence. The social environment may modify the way cratic pursuits are gratified to a certain extent, but they remain largely independent of it. According to Adler, a sense of inferiority as a universal human experience is common to all people and appears already in the processes of socialization. In other words, Adler regarded the sense of inferiority as a source and driving force behind the will to power, while Witwicki believed that a sense of inferiority appears when the possibility of achieving a sense of power is blocked. Or, in yet other words: socialization affects the ways of dealing with inferiority, or the ways in which cratic desires are realised.

Thirdly, an analysis of the ways of dealing with the sense of inferiority led Adler to believe that there is no one style of achieving the sense of power that is common to all people. Everyone develops a unique style on which the entire structure of their personality and the type of relationships they develop with the environment are based. Adler calls it *Lebensstil*, and it is usually not consciously identified by the individual, although it can and should be identified by a psychologist during the therapeutic process. He also assumed that one's *Lebensstil* may be inadequate and may lead to various intra- and interpersonal problems. Adler purposefully refrained from constructing of a typology of personalities; he was aware that *Lebensstil* is always unique, different for every human being. Witwicki did not use a distinctive typology of personalities either, but distinguished four main ways of achieving power (elevating oneself/others, diminishing oneself/others). As Estera Markinówna (1935) noticed quite recently, a vast majority of cases described by Adler could be assigned to one of the four ways of achieving power discussed by Witwicki.

Naturally, Adler's theory differs from Witwicki's on several important points. First of all, Adler took care to extensively explain the causes of diversity in people's lifestyles, referring to complex family and social factors. Witwicki merely mentioned four types of cratic desires, without explaining their background. In addition, Adler performed a comprehensive analysis of pro-social and anti-social behaviour, introducing the notion of a sense of community (*Gemeinschaftsgefühl*). In his opinion, only this strategy of dealing with a sense of inferiority and achieving a sense of power was desirable and acceptable, and contributed to building harmonious social relationships. To use Witwicki's terminology, this is characterised by the elevation of oneself or of others. Identification of anti-social types and indication of ways to modify them began the highly influential therapeutic school in the spirit of Adler's individual psychology. An educational programme was successfully developed and popularised in many countries around the world. Unfortunately, Witwicki's theory has never been introduced into the social practice of psychologists and educators. Moreover, it has remained a Polish theory, without reference to the achievements of other psychologists around the world. It is rather surprising that Witwicki never started a dialogue with Adler and was not interested in Adler's theory.

3 'Wiara oświeconych', 'Dobra Nowina wg Mateusza i Marka' and the Dorpat School of the Psychology of Religion

'Wiara oświeconych' ('The Faith of the Enlightened'), widely known to Polish readers, presents Witwicki's original research into the coherence of beliefs (religious, ethical, etc.). This is probably the earliest Polish psychological dissertation which discusses the problem of not only beliefs, but also the problem of religious experience—or at least some of its aspects. Witwicki—despite his atheist worldview—was keenly interested in religion and religious beliefs, especially in the relationship between logic and faith.

Witwicki also wrote the less known 'Dobra Nowina wg Mateusza i Marka' ('The Good News According to Matthew and Mark') (Witwicki 1958), in which he presented a cratic interpretation of the life and work of Jesus Christ. Both studies are among the very few examples of research into the field of psychology of religion in the Lvov-Warsaw School. Their final shape was undoubtedly influenced by Witwicki's personal childhood experience and family life, including a rather strict religious atmosphere at home, which he later mentioned on a number of occasions. In spite of being affected by Witwicki's personal worldview, they present the results of his empirical research and constitute an important contribution not only to Witwicki's scientific work, but to the Polish psychology of religion as a whole.

3.1 The Problem of Suppositions and Cognitive Dissonance

The theory of suppositions was developed mainly by Alexius Meinong, a student of Brentano, who claimed that there is a type of judgment which does not need to be related to any conviction about the actual state of affairs (Meinong 1910). Suppositions do not explicitly describe any existing state of affairs, and may be held or expressed without any actual conviction or certainty about the reality they refer to. In the Lvov-Warsaw School, the notion of supposition was analysed by Twardowski (1898, 1924), Ajdukiewicz (1965, vol. 1) and Witwicki (1939/1959). Twardowski distinguished expressed (actual) judgments and conceived (potential) judgments. While according to Ajdukiewicz there are logical and psychological judgments, only the latter may be accepted or rejected, and only psychological judgments, lending themselves to acceptance or rejection, were considered a supposition (cf. Jadczak 1979: 37).

Witwicki, on the other hand, carried out an in-depth analysis of suppositions on the theoretical and empirical level. He found that every judgment or conviction about the world was based on some kind of representation; that such judgements may be deemed true or false; and that the ontological principle of contradiction applied to them as well (Witwicki 1926/1962). Nevertheless, thinking about the world as a mental activity may result in such mental products as suppositions, i.e. a kind of speculation or an apparent conviction. Just like Meinong, Witwicki treated supposition as an 'intermediate state between a conviction and a 'mental image' (Witwicki 1939/1959: 28). A supposition as a 'speculation' may then be very close to a conviction or very close to a mental image, although it belongs to neither category. Suppositions are the mental products of mental acts whose object is indefinite, vague, or whose actual existence and features cannot be determined. They are very popular in poetry, aesthetic impressions, fine literature, dreams, religion, reported speech and modals of certainty (Witwicki 1926/1962).

Witwicki (1939/1959) used introspective reports in his research. The results led him to conclude that thinking about religion, ethics or aesthetics is often based on suppositions. The persons participating in his experiment made statements that appeared untruthful or false from the perspective of logic. The most interesting element of his study was suspension of the principles of logic with respect to convictions regarding God or religious doctrines. The participants had no problem passing ethical judgments on imaginary persons who violated moral standards in a fictional story. However, when they were requested to judge the same category of deeds by persons who are considered morally impeccable (such as saints or Biblical characters), a vast majority of respondents introduced and used a number of suppositions into their own system of convictions in order to uphold their earlier judgment. In other words, to avoid a conflict between two judgements or convictions, they used new convictions or suppositions ('killing your own son instead of an offender is wrong, but if God does it, it is right because…').

The most important of Witwicki's findings discussed in 'The Faith of the Enlightened' is his discovery of the suspension of logical thinking with respect to religious or ethical beliefs in the face of an acknowledged contradiction. It was a revolutionary experiment at that time, not only because it was conducted within a similar methodological paradigm as that used by the Dorpat School (Wulff 1985), where experiments employing introspection were used in pioneering studies on religious experience (Girgensohn 1921; Gruehn 1924), but mainly on account of the fact that—already in 1939—Witwicki was able to prove the fundamental cognitive mechanism of reducing psychological tension arising from a conflict of beliefs (Rzepa 1997). In 1939, he explicitly wrote:

> A deeply watchful man avoids any contradictory convictions; upon acknowledging such a contradiction, he feels anxious and believes he is in error. He is prepared to adhere to one out of two contradictory convictions, although at times he may find it difficult to choose the right one. It is particularly hard for a reasonable person to experience two contradictory convictions simultaneously and consciously. (Witwicki 1939/1959: 65)

His findings suggested that the psychological principle of contradiction was not followed by man as rigorously as the logical principle of contradiction. It should be added that, following Jan Łukasiewicz (Łukasiewicz 1909/1910), Witwicki distinguished the principles of ontological, logical and psychological contradiction, introducing this division even in his 'Handbook of Psychology' (Witwicki 1926/1962; Rzepa 1997).

This psychological mechanism was described in detail by Festinger 18 years later, in 1957, as a well-developed and empirically substantiated theory of cognitive dissonance. According to this theory, the psychical tension caused by the presence of mutually contradictory beliefs is reduced, for example, by their modification or removal (Festinger 1957/2007). In Witwicki's findings, they are 'transferred' outside logical rules, and as such they can still exist in man's mental space. Unfortunately, throughout the text Witwicki expresses a more or less negative attitude to religious beliefs as such, comparing them even to insincere or childish beliefs which do not become an educated person. His important scientific discovery was thus subordinated to his private worldview, which generated two serious consequences. Firstly, it hindered a more in-depth analysis of the psychological and logical principle of contradiction (Rzepa 1997). Secondly, it made it difficult, if not impossible, to understand the nature of religious beliefs, or at least some of them. However, as emphasised by Teresa Rzepa, the Polish expert on Witwicki's psychological work, in 'The Faith of the Enlightened' 'Witwicki presented himself as [...] a pioneer of the theory of cognitive dissonance' (Rzepa 1997: 129). Cognitive dissonance is currently one of the most popular and empirically confirmed psychological theories, even though it was 'discovered' by Witwicki and Festinger in the course of their research on religious beliefs (Festinger et al. 1956/2012).

Another very important (although controversial and underestimated) of Witwicki's achievements in the psychology of religion was his cratic interpretation of Jesus Christ's personality, presented in 'The Good News According to Matthew and Mark'. In this study, he ascribes cratic motives to the religious activity and awareness of Jesus, and attributes to him a desire for superiority and domination over others. The cratic tendencies Witwicki attributes to Jesus are primarily negative—they consist in seeking to elevate himself and diminish (!) others.

Although there are considerable limitations and weaknesses to his analysis, it is nevertheless very interesting, and in my opinion testifies to Witwicki's extraordinary intuition. The pursuit of power was, in fact, one of the most important factors affecting social relationships in the ancient world, and played an important role in the experience of the *sacrum*. It can be seen in both the Old and the New Testament, and in many other written sources of antiquity (Citlak 2016b, 2019; Johnson and Earle 1987).

The 'master-servant', 'stronger-weaker', 'ruler-subject' perspective is so emphatically present there that it can be seen even in a superficial reading. Witwicki attributes cratic desires to religious figures, such as Moses, John the Baptist, Old Testament prophets, and even Catherine of Siena and Francis of Assisi (Witwicki 1927). This suggests that for Witwicki cratic considerations played an important role in religious experience, although he did not perform any in-depth analyses or venture beyond general statements and conclusions on the subject. However, his propositions were of extraordinary importance and could serve as a point of departure for in-depth empirical research on the structure and specific nature of religious experience, which was keenly studied in psychology not only in his times, but is still so today.

Research on religious experience was conducted at the beginning of the twentieth century in both the USA and Europe, with particular success among representatives of the so-called Dorpat School of the Psychology of Religion (Wulff 1985). As in the case of research on the persuit for the sense of power, Witwicki was aware of Adler's work, and familiar with research on religious experience conducted by K. Girgensohn, founder of the Dorpat School, whom he mentions in 'The Faith of the Enlightened'.

Unfortunately, he only mentions him without making any specific reference, and compares his own findings with the results of works by Karl Girgensohn (1921), Werner Gruehn (1924), and other representatives of the School. It was, indisputably, a very impressive tradition of research. Girgensohn had a large group of followers whose achievements in this field belong to the classics of the psychology of religion, and who set the standards of research in this discipline for many decades to come.

There is no room for a detailed description of the Dorpat School here, so I will only mention that it brought together psychologists from all over Europe, closely cooperating with 'Archiv für Religionspsychologie', the most influential psychological journal in this field, where ground-breaking empirical achievements were published on a regular basis. In their research, they employed a strictly defined methodology in the paradigm of systematic introspection (Allik 2006). A great number of studies were used to verify philosophical statements about the nature of religious experience, like that of F. Schleiermacher (religion as a state of affective dependence), or R. Otto (experience of the *sacrum*) (Wulff 1997). Philosophical tradition was very skilfully combined here with the empirical tradition of psychology, which was—in my opinion—also possible in Witwicki's research.

Surprisingly, however, Witwicki never consolidated the results of his work with the achievements of Dorpat psychologists. Parallels between them were not just limited to similar methodology, philosophical tradition, or field of study. The benefits of such consolidation would result primarily, both then and now, from the complementarity of their findings. This applies both to religious beliefs and religious thinking, and to a sense of power inherent to religious experience.

> The achievements of Dorpat psychologists (to use a collective term) proved beyond any doubt that religious beliefs which arise from religious

> experience are not necessarily suppositions which evade the principles of logic and common sense. Religious beliefs are usually based on intuitive thinking; they are an attempt at capturing an experience that is not always rational [...] The findings of Girgensohn and Gruehn make us look at the problem of religious beliefs [...] from a wider perspective, one the is not limited to logical *versus* illogical, sincere *versus* pretend, or believed *versus* supposed. Such wider perspective may help us avoid seeing only the conflictual nature of religious beliefs and Witwicki's anti-clerical attitude. Moreover, there seems to be some convergence between Girgensohn's findings about the ego experiencing omnipotence and power in the act of unification with an indefinite religious Entity/Being on the one hand, and the cratic motives in life and work of biblical figures like Jesus Christ, John the Baptist or Moses postulated by Witwicki on the other. I believe their studies may, to a certain extent, be considered complementary. (Citlak 2016a: 295)

Witwicki's psychology of religion is also an example of the rarely encountered two-track psychology of religion, whose foundations were laid already by Wundt at the turn of the nineteenth and twentieth centuries. Wundt, as is well known, was the author and proponent of the 'experimental' paradigm in psychology, the counterpart of which, in a way, was Witwicki's research on religious beliefs with the application of systematic introspection. Within this framework, a person could be examined based on the empirical material they have provided.

Apart from experimental research, Wundt also saw the need for psychological research of a more qualitative nature. He referred to the *Völkerpsychologie* tradition, already popular in the German literature, discussed by W. von Humboldt in his writings and, after 1860, in the journal 'Zeitschrift für Völkerpsychologie und Sprachwissenschaft'. It focused on research into language, custom, myth, and religion from a historical and comparative perspective. Its goal was to trace the evolution of mental processes and identify the laws of its development, and to create a larger interpretative framework for experimental research (Wundt 1912). Witwicki's works on ancient documents such as the gospels, books of the Old Testament, or Plato's 'Dialogues' fit perfectly in this paradigm formed by the father of psychology.

However, it should be emphasised that in both cases the feature that distinguishes Witwicki's two-track research is primarily an affinity with the descriptive psychology advocated by F. Brentano, i.e. *deskriptive* and *verstehende Psychologie*. Of course, 'The Faith of the Enlightened' is not a typical example of experimental psychology, although it has much more in common with it than with historical-cultural psychology. Unfortunately, also in this case, Witwicki did not attempt to compare or consolidate his results with the work of Dorpat psychologists.

4 Cratic Portraits and Psychobiographies of Socrates and Jesus Christ

The theory of cratism was used by Witwicki to develop two psychobiographies: that of Jesus of Nazareth and of Socrates. I will only refer to the most important theses, in order to avoid repeating many of Witwicki's propositions. The psychological portrait of Socrates was developed on the basis of Plato's 'Dialogues' and not only implied in the translation, but also explicitly referred to in comments on the translated texts.

The earliest references to cratic elements in Socrates' personality are found in commentaries on the Symposium of 1909 (Witwicki 1909). The image of the Greek philosopher was gradually complemented with subsequent translations and comments, which made it a dispersed work, although entirely in line with assumptions of the theory of cratism. Socrates is seen in this interpretation as motivated by the desire to elevate himself and to elevate others, although diminishment of others plays an important role in his behaviour as well, which is particularly visible when he engages in polemics with opponents of his own worldview. A sense of power is achieved by Socrates not only by acquiring knowledge and learning truth about the world, but above all by an intellectual and mental superiority over those around him (Rzepa 2002). The psychological image of Christ appears to be quite similar: Witwicki attributes to him the desire to elevate himself and to diminish others.

However, the latter motivational mechanism plays a very important role here and is definitely stronger than in the case of Socrates.

Many of Christ's deeds (even miraculous healings) and teachings are interpreted as a desire to diminish (!) and subordinate other people to himself. Witwicki projects a psychological (cratic) paradigm on the sense of divinity and messiahship of Jesus, which he believes to be nothing but a desire to rule over others and to elevate himself. In his analysis of the 'Sermon on the Mount', Witwicki even came to the conclusion that Jesus deliberately evoked feelings of guilt and sin in his listeners in order to become their saviour. So, while the psychobiography of Socrates appears to be balanced and deserving appreciation, that of Jesus raises a number of considerable objections (Citlak 2015; Smereka 1961).

However, without going into a dispute about the strength of Witwicki's arguments, the exceptional originality his work should certainly be emphasised. His psychobiographies are unique for several reasons. Firstly, Witwicki took up the difficult task of creating a psychological portrait of ancient figures—a very complicated one considering the specific nature of written sources available to him, and in view of the historical and cultural gap he had to face. In both cases, Witwicki also had to wade through the linguistic thickets formed by the network of cultural and social meanings contained in the ancient Greek language of Plato's 'Dialogues' and Biblical gospels.

Secondly, these psychobiographies are entirely independent from the earliest psychobiographies ever written, based on the assumptions of Freud's psychoanalysis. Moreover, it was the analysis of ancient sources that led Witwicki to develop his theory of cratism and to create (or, as the author himself wanted: to re-create) the cratic portrait of the great masters of antiquity. In my opinion, this is a truly distinctive aspect of Witwicki's investigations, who did not submit to the psychoanalytic 'fashion', but was able to develop his own theoretical perspective and indicate its practical applications. In addition, psychobiography still remains largely dominated by the psychoanalytic tradition, which, while criticised, is rarely contrasted with any alternative theories or interpretations (Pawelec 2004).

Thirdly, we can now boldly consider them as some of the earliest psychobiographies in the world (especially regarding the portrait of Socrates), as well as the earliest non-psychoanalytic biographies (see Runyan 1982). One of the earliest documented psychobiographies is

Freud's psychoanalytical interpretation of the life and work of Leonardo da Vinci, which is a continuation of his earlier publication of 1907. Thus, Witwicki could be seen among the top psychobiographers, along with such figures as Freud (1910, 1939/1994), Erikson (1958, 1969), and others. Of course, this did not happen because Witwicki never presented the results of his work to a wider audience, outside of the scientific community in Poland.

Fourthly, Witwicki's works discussed here indicate one of the most important mechanisms governing human life, i.e. cratic desires and motives. There are similarities between his theory and many later theories developed in psychology, by Adler as well as others. Significant analogies can be seen, for example, in the works of Theodor Adorno on authoritarian personality and on sadomasochism (Adorno et al. 1950; Adorno 1975/2010). Other examples include the model of interpersonal relationships developed by the American psychologist Timothy Leary (Leary 1957), combining the dimensions of domination *vs* submission, and friendship *vs* hostility. Certain analogies may also be found in the theory proposed by Charles McClintock (McClintock 1972, 1978), where social relationships are categorised according to a person's desire to maximise or minimise their own or someone else's benefits.

However, there is more to Witwicki's works. His theory was developed to a large extent on the basis of an analysis of social relationships which can be deducted from ancient documents. The figure of Socrates and the constellations of cratic desires in his Grecian world are different than in the case of Jesus Christ, who was a representative of Semitic culture. It is very likely that there is some connection between the structure of a society and the personality of its members, as was discussed in the Lvov-Warsaw School by Andrzej Lewicki (Lewicki 1960) and Stanisław Ossowski (Ossowski 1967/2000).

This relationship appears to be quite obvious to Witwicki, even though he did not discuss it explicitly in his writings. He made an important contribution to the tradition of psychobiography, marked in the same moment and a bit later by representatives of the Annales School in France (Grabski 2011).

Finally, it must be emphasised that Witwicki's psychobiographies are very valuable as well, demonstrating that cratic desires could have played a pivotal role in the organization of social life and social relationships in the ancient world. Interestingly, recent Polish linguistic studies on ancient (religious) documents have confirmed this thesis quite clearly (Citlak 2016c), and Witwicki seems to have identified one of the important psychological mechanisms operating in the ancient world.

Although a desire for power appears in both historical and anthropological literature discussing ancient culture, it has never been explicitly referred to the psychological perspective in the way proposed by Witwicki. Unfortunately, also in this case he did not enter into a dialogue with the relevant branch of psychology, nor refer his results to the psychobiographies available to him at that time (e.g. Schweitzer 1913; Hall 1917; see Runyan 1982).

The psychological inquiries of Witwicki can be considered the most original works of Polish psychology in the first half of the twentieth century. They were close to the important and widely-recognised theoretical and empirical achievements in the world of scientific psychology at the time. His works are characterised by a distinctively interdisciplinary nature, discussing issues in cognitive psychology, the psychology of religion, the psychology of personality, and historical psychology. The fact that the author never made an attempt to consolidate them with the achievements of other psychologists is, of course, surprising and difficult to explain. However, I would like to emphasise that many psychological theses he proposed still have a high explanatory value, and I believe they have a predictive value as well. It is difficult to design psychological research based solely on Witwicki's theories, but they may be an important inspiration, especially if they are associated with the related theories whose empirical status has been verified and confirmed.

References

Adler, A. (1907). *Studie über die Minderwertigkeit von Organen*. Berlin and Wien: Urban & Schwarzenberg.

Adler, A. (1912). *Über den nervösen Charakter. Grundzüge einer vergleichenden Individualpsychologie und Psychotherapie*. Wiesbaden: Verlag von J.F. Bergmann.

Adler, A. (1920). *Praxis und Theorie der Individualpsychologie*. München: Verlag von J.F. Bergmann.

Adorno, T. W. (1975). *Studies in the authoritarian personality*. Frankfurt am Main: Suhrkamp Verlag.

Adorno, T. W., Frenkel-Brunswik, E., Levinson, D., & Sanford, R. N. (1950). *The authoritarian personality*. New York: Harper & Row.

Ajdukiewicz, K. (1965). *Język i poznanie. Wybór pism z lat 1945–1963* [Language and Cognition. Selected Works 1945–1963]. Warszawa: PWN.

Allik, J. (2006). History of experimental psychology from an Estonian perspective. *Psychological Research, 71,* 618–625.

Citlak, A. (2015). Psychobiography of Jesus Christ in light of kratism theory. *Journal for Perspectives of Economic Political and Social Integration: Journal for Mental Changes, 21*(1–2), 155–184.

Citlak, A. (2016a). The Lvov-Warsaw school—The forgotten tradition of historical psychology. *History of Psychology, 19*(2), 105–124.

Citlak, A. (2016b). O psychologii religii w szkole lwowsko-warszawskiej i szkole dorpackiej. In J. Bobryk (Ed.), *Język, wartości, działania. Szkoła lwowsko-warszawska a wybrane problemy psychologii, semiotyki i filozofii* (pp. 271–298). Warszawa: Instytut Psychologii PAN.

Citlak, A. (2016c). *Relacje społeczne świata antycznego w świetle teorii kratyzmu. Psychologia historyczna w szkole lwowsko-warszawskiej*. Warszawa: Instytut Psychologii PAN.

Citlak, A. (2019, in press). The concept of "cratism" and "heteropathic feelings" in the psychobiography of Jesus from Nazareth (psychobiography in Lvov-Warsaw School). In C. H. Meyer & Z. Kovary (Eds.), *New Trends in Psychobiography*. Springer.

Dosse, F. (1994). *New history in France: The Triumph of the Annales*. Illinois: University of Illinois Press.

Erikson, E. H. (1958). *Young man luther: A study in psychoanalysis and history*. New York: W. W. Norton.

Erikson, E. H. (1969). *Gandhi's truth: On the origins of militant nonviolence*. New York: W. W. Norton.

Festinger, L. (1957). *A theory of cognitive dissonance*. Stanford: Stanford University Press.

Festinger, L., Riecken, H., & Schachter, S. (1956). *When prophecy fails*. Minneapolis: University Minnesota Press.

Freud, Z. (1910). *Eine Kindheitserinnerung des Leonardo da Vinci*. Leipzig and Wien: Franz Deutike.

Freud, S. (1939). *Der Mann Moses und die monotheistische Religion*. Amsterdam: Verlag Allert de Lange.
Girgensohn, K. (1921). *Der seelisches Aufbau des religiösen Erlebens. Eine religionspsychologische Untersuchung auf experimentaler Grundlage*. Leipzig: Hirzel.
Grabski, A. F. (2011). *Dzieje historiografii*. Poznań: Wydawnictwo Poznańskie.
Gruehn, W. (1924). *Das Werterlebnis. Eine Religionspsychologische Studie auf experimenteller Grundlage*. Leipzig: Hirzel.
Hall, S. (1917). *Jesus the christ in the light of psychology*. New York: Doubleday.
Jadczak, R. (1979). Teoria supozycji wg Władysława Witwickiego [Theory of suppositions according to Władysław Witwicki]. *Acta Universitatis Nicolai Copernici, 4*(103), 35–45.
Jadczak, R. (1981). Teoria kratyzmu Władysława Witwickiego. *Acta Universitatis Nicolai Copernici, 5*(121), 25–40.
Johnson, A. W., & Earle, T. (1987). *The evolution of human societies: From foraging group to Agrarian State*. Stanford, CA: Stanford University Press.
Leary, T. (1957). *Interpersonal Diagnosis of Personality*. New York: Ronald.
Lewicki, A. (1960). *Procesy poznawcze i orientacja w otoczeniu*. Warszawa: PWN.
Łukasiewicz, J. (1909). Über den Satz des Widerspruchs bei Aristoteles. *Bulletin International de l'Académie des Sciences de Cracovie, Classe de Philosophie et d'Historie*, 15–38.
Markinówna, E. (1935). Psychologia dążenia do mocy. Zestawienie poglądów Witwickiego i Adlera. *Kwartalnik Psychologiczny, 7*, 329–340.
McClintock, C. (1972). Social motivation—A set of propositions. *Behavioral Science, 17*, 438–454.
McClintock, C. (1978). Social values: Their definition, measurement and development. *Journal of Research and Development in Education, 12*, 121–137.
Meinong, A. (1910). *Über Annahmen*. Leipzig: Barth.
Nowicki, A. (1982). *Witwicki*. Warszawa: Wiedza Powszechna.
Ossowski, S. (1967/2000). *Z zagadnień psychologii społecznej*. Warszawa: PWN.
Pawelec, T. (2004). *Dzieje i nieświadomość. Założenia teoretyczne i praktyka badawcza psychohistorii*. Katowice: Wydawnictwo Uniwersytetu Śląskiego.
Runyan, W. (1982). *Life historie and psychobiography*. Oxford: Oxford University Press.
Rzepa, T. (1991). *Psychologia Władysława Witwickiego*. Poznań: Wydawnictwo Naukowe UAM.

Rzepa, T. (1997). Dlaczego Władysław Witwicki nie odkrył zjawiska dysonansu poznawczego? *Przegląd Filozoficzny, 1*(21), 79–93.
Rzepa, T. (2002). *O interpretowaniu psychologicznym w kręgu szkoły lwowsko–warszawskiej*. Warszawa: Polskie Towarzystwo Semiotyczne.
Schweitzer, A. (1913). *Die psychiatrische Beurteilung Jesu. Darstellung und Kritik*. Tübingen: Mohr Siebeck.
Smereka, W. (1961). Prof. Witwicki jako tłumacz ewangelii. *Ruch Biblijny i Liturgiczny, 14*(3–4), 114–120.
Twardowski, K. (1898). Wyobrażenia i pojęcia. In K. Twardowski (Ed.),. *Wybrane prace filozoficzne* (pp. 114–197). Warszawa: PWN. (1965).
Twardowski, K. (1924). O istocie pojęć. In K. Twardowski (Ed.), *Wybrane prace filozoficzne* (pp. 292–312). Warszawa: PWN. (1965).
Witwicki, W. (1900). Analiza psychologiczna ambicji. *Przegląd Filozoficzny, 3,* 26–49.
Witwicki, W. (1907). Z psychologii stosunków osobistych. *Przegląd Filozoficzny, 10,* 531–537.
Witwicki, W. (1909). Wprowadzenie, komentarze, w: Platon. *Uczta*. Lwów: Księgarnia Polska B. Połonieckiego.
Witwicki, W. (1926/1962). *Psychologia*. Tom 1. Warszawa: PWN.
Witwicki, W. (1927/1963). *Psychologia*. Tom 2. Warszawa: PWN.
Witwicki, W. (1939). *La foi des eclaires*. Paris: Alcan. Polish edition: 1959, *Wiara oświeconych*. Warszawa: PWN.
Witwicki, W. (1958). *Dobra Nowina wg Mateusza i Marka*. Warszawa: PWN.
Wulff, D. (1985). Experimental introspection and religious experience: The dorpat school of religious psychology. *Journal of the History of the Behavioral Sciences, 21,* 131–150.
Wulff, D. (1997). *The psychology of religion: Classic and contemporary*. New York: Wiley.
Wundt, W. (1912). *Elemente der Völkerpsychologie*. Leipzig: Kröner.

Part III

Logic and Methodology

10

Pragmatic Rationalism and Pragmatic Nominalism in the Lvov-Warsaw School

Witold Marciszewski

1 Introduction

According to Francis Bacon, 'The understanding must not [...] be supplied with wings, but rather hung with weights, to keep it from leaping and flying' (Bacon 2011, Aphorism CIV: 97). Let us update Bacon's metaphor of wings and weights by speaking instead about gears and brakes in a motorcar. Moreover, let us complete this picture with the fact that both devices are necessary, but in very different ways. When driving, we need the driving mechanism constantly, while the brakes only to pull up or slow down.

This comparison reflects the roles and contributions of two competing philosophies of science. These are: pragmatic 'rationalism' as the gearing system, and 'nominalism' which attempts to operate as the

W. Marciszewski (✉)
Foundation for Computer Science,
Logic and Formalized Mathematics, Warsaw, Poland
e-mail: witmar@calculemus.org

A. Drabarek et al. (eds.), *Interdisciplinary Investigations into the Lvov-Warsaw School*, History of Analytic Philosophy,
https://doi.org/10.1007/978-3-030-24486-6_10

breaking system. Unfortunately, these attempts prove to be excessive and greatly detrimental to the progress of science and philosophy (if they were to succeed).

Nevertheless, nominalism should be appreciated for emphasising the role of sensory perceptions concerning physical objects; that is, objects given to us in our everyday experience, being the starting point for sciences on their way to more abstract regions. This emphasis on the import of the everyday experience of physical things (Lat. res) is what justifies both keeping the name 'reism', and freeing it from its aversion towards abstract objects, since these are indispensable in any scientific inquiry.

Such a bold attempt at reforming reism in along the lines of pragmatism was undertaken by the eminent mathematician and logician Andrzej Grzegorczyk (cf. Grzegorczyk 1997: 12) called it 'reism in a liberal sense' (see paragraph 2.3). Since reism was understood by Tadeusz Kotarbiński as a kind of nominalism, while Grzegorczyk's liberalisation was motivated by pragmatic considerations (the problem with efficiency of scientific inquiry), in this paper Grzegorczyk's proposal will be called 'pragmatic nominalism' (an adjective as good as 'liberal' or 'moderate'; cf. Marciszewski 2012).

Pragmatic rationalism (the view held by the author of this paper) is focused on the processes of creative 'abstraction', and on abstract entities as either constructed or discovered in the course of those processes (the interplay between construction and discovery is too subtle an issue to be dealt with in the present paper). On the other hand, nominalism pays careful attention to dangers of possibly excessive abstraction. Therefore, it looks for measures to prevent possible cognitive damage. Thus, e.g., Stanisław Leśniewski tried to take nominalist measures against the antinomies of the abstract set theory.

The concern for 'abstract entities', or 'abstracts' (for short), is common to rationalism and Platonism, hence the two terms happen to be used interchangeably. The present paper does not distance itself from such use, but with the proviso that abstracts are not treated as if perceived—as Plato imagined—in a direct and infallible vision, but as a result of a complex cognitive process, not immune to error. On account of the lack of full certainty, existential propositions about abstracts should be tested against their applications: to check if they 'work' in explanations, predictions, etc. The requirement of such practical application is what justifies the use of the adjective 'pragmatic'.

The survey of opinions and arguments presented in this paper is far from complete. The main emphasis is placed on the Lvov-Warsaw School, mainly on Leśniewski, Kotarbiński, Janina Kotarbińska (as nominalists), and Kazimierz Ajdukiewicz, Alfred Tarski, and Roman Suszko (as pragmatists). This is motivated by the School's highly significant contributions to international debates.

Obviously, the School's contributions should be analysed within a broader international context, for instance that of Nelson Goodman's defence of nominalism, or the pragmatist offensive of Willard V. O. Quine. The moral of the story told in this paper is to the effect that abstraction is an enormous driving force towards the intellectual progress of mankind.

It is worth noting that the faculty of abstraction, an astonishing feature of the human mind, presents a considerable challenge for the project of artificial intelligence. It seems very questionable whether a machine can ever obtain a power of abstraction to match humans in such brilliant achievements as the abstract set theory, relativity, quanta, etc. This is why pragmatic rationalism, focused on the study of abstracts and abstraction, is such a focal point in the philosophical landscape of our times (cf. Marciszewski 2017).

2 On How to Do Science: Controversies About Nominalism in the Lvov-Warsaw School

2.1 What Does It Mean to Be Indispensable? Ockham's Maxim in a Modern Interpretation

The modern contention between nominalism and its opponent, conventionally named Platonism (without claim to historical accuracy), has little to do with the medieval controversy about universals. The main difference consists in the fact that in contemporary debate the question about the existence of some types of entities is inseparable from the question about scientific utility: which position is more useful and conducive to the development of science?

This is due to the modern perception of science as evolving infinitely, contrary to our medieval ancestors' belief that the whole available science had been achieved in antiquity. The task of posterity was supposed to consist in simply reconstructing the achievements of ancient thinkers and commenting on them for the benefit of future generations.

Though the issue of universals still remains relevant in the modern debate, they are seen as just one among various types of abstract objects: classes (sets), numbers, functions, objects of geometry, states of affairs, etc. Some of them, e.g. natural numbers, can be treated either as universals, i.e. classes of classes (e.g. number two as a class of all pairs of objects) or as individuals, for instance, in the domain of Giuseppe Peano arithmetic.

After the concept of an abstract entity has been properly defined, it plays the main role in distinguishing between parties to the debate. Rationalism admits in a scientific theory statements which assert the existence of abstract entities, while nominalism does not, charging such assertions with being either false or absurd (depending on the version of nominalism).

Among the legacies of medieval discussions, there is the famous maxim by William of Ockham: entia non sunt multiplicanda praetor necessitatem, meaning 'entities are not to be multiplied beyond necessity'. We may be disappointed with the triviality of this statement. Economy is a virtue in handling anything: nothing is to be done unless it is reasonably necessary, and this refers to multiplying entities as well. It would be equally right to say that entities should not be reduced beyond necessity either.

However, this postulate of economy proves fertile if taken not as rule of behaviour, but as the demand to decide which means are indispensable to reach a desired goal. If our goal is defined as the progress of science, then we look for suitable means. For instance, according to Ajdukiewicz's (1965: 2–3) programme 'for an empirical methodology of science', we should learn from the history of science which concepts and methods are indispensable for scientific progress.

This is in accord with Quine's famous idea that indispensability is what characterises in particular the abstract concepts of mathematics. This entails the existence of abstract entities, which these mathematical concepts refer to. Should anybody need a more accurate account of what 'indispensable' or 'unnecessary' means, let them consider the following explication (cf. Colyvan 1998).

Definition 1

An entity is 'dispensable' in a theory if there exists a modification of that theory resulting in a second theory, functionally equivalent to the first, in which the entity in question is neither mentioned nor predicted. Furthermore, the second theory must be preferable to the first. Otherwise, the entity in question is acknowledged as 'indispensable' in scientific research.

In mathematics, we frequently find assertions prefixed with the phrase 'for every *x*, there exists *y* such that...'. Their truth implies that there are objects being values of the variable *y*. If *y* belongs, say, to the category of functions, this means that functions exist. Since functions are abstract entities, it follows that there exist abstract entities. Hence, Quine's maxim: 'to be is to be the value of a bound variable'.

Kotarbiński, the leader of the nominalist orientation in the Lvov-Warsaw School, sees the problem otherwise. He allows of existential quantification over variables for semantic types other than individuals, but in such cases one should stick to the following crucial rule: the meaning of 'exists' as a reading of the existential quantifier is essentially different from the meaning of 'exists' in the proper language of philosophy (cf. Kotarbiński 1957: 159–160).

Any object existing in the proper sense must be—according to the nominalist approach—an individual. Kotarbiński conceived individuals restrictively as physical tridimensional objects (solids). He believed that such point makes us free from any philosophical troubles or puzzles of existence. However, a serious puzzle arises when reistic nominalism is confronted with modern physics.

2.2 The Puzzle of Existence: What Exists in the Domain of Physics?

In his manifesto of reism, Kotarbiński (1979) put forward a puzzling view denying the existence of waves in the physical universe. His remark is directly concerned with quanta, but the argument applies to any wave phenomena, those of classical physics as well.

> It is not possible to accept such an interpretation of experimental data in which a particle of a physical body, and hence a small object, would have, under certain conditions, to be identical with a wave. The noun 'wave' is a special case of the general term 'process', and as such, from the reistic point of view, yields nonsense when substituted for the name of a thing in the ultimate formulation. (Kotarbiński 1979: 48)

Were this argument valid, then even the expression 'waves of water' should be regarded as nonsense, though it refers to a physical, directly observable phenomenon. The statement 'I can see a wave' could not account for sensory perception, and would remain inexpressible unless the observer articulated his perception in terms of moving particles. However, he cannot see individual molecules of water; what he sees is just the movement of a mass of waves. Again, 'movement', according to reism, would be another nonsensical sequence of letters which does not denote anything.

A way out might be sought in distinguishing two roles of the terms 'exists', 'there is' etc. There was a famous controversy among philosophers as to whether 'exists' is a 'predicate' or belongs to another category of expressions. In symbolic logic, that other category is defined as 'variable-binding operators', which include the 'existential quantifier'.

In the logical and ontological controversy about how the word 'exists' should be interpreted, most eminent authors were involved. Gottlob Frege and Bertrand Russell rejected the idea of existence as a property denoted by a predicate, while Alexius Meinong and his followers made just such a claim and defended it vigorously (cf. Nelson 2012).

What Kotarbiński meant by the above passage might be as follows. 'Exist' as a predicate is applicable to bodies alone, including subatomic particles, while variables ranging over abstracts (including physical fields) may be bound by the quantifier conventionally called existential, but having nothing to do with actual existence.

This issue might be cleared by introducing a distinction between 'referential' and 'substitutional' interpretation of quantifiers. The former is characteristic of the classical quantification logic. On the other hand, under substitutional interpretation the formula ∃xA(x) might be read as: 'for some substitution instances of A(x) this formula is true', instead of the classical referential reading: 'there is at least one thing which satisfies A(x)'.

Quantification in Leśniewski's ontology is certainly not referential, and therefore it does not entail ontological commitment in Quine's sense. This feature can be seen in the fact that in Leśniewski's system both of the following assertions hold:

$$(*)\ \ -ex(Pegasus)\ \ hence:\ \ (**)\ \exists y - ex(y).$$

The turned 'E' means here the existential quantifier in non-referential sense; 'ex' means the existential referential quantifier (in Leśniewski's idiom), while the dash renders negation. The formula (**) is intended to mean: for some names, there does not exist any referent, e.g. the referent of the name 'Pegasus'. By analogy: for some names of classes, e.g. 'a pair' (two-member class), there do not exist 'classes' so named.

Whether Leśniewski's approach is, in fact, substitutional or, possibly, of yet another kind, is a much debatable issue. Experts' opinions are divided. The matter is crucial for understanding Leśniewski's and Kotarbiński's nominalism, but too involved to be discussed in the present context. Brief but illuminating suggestions are offered by Peter Simons (1992: 279), and Guido Küng (1981: 170–174).

In his essay (1979) and lectures on the history of logic (1957: 156–159), Kotarbiński sketched out the nominalist guidelines for doing mathematics, and, thereby, its application in sciences. He challenged the standard use of the same existential quantifier for both concrete physical things and abstract objects, and postulated recourse to Leśniewski's logic instead. However, he seemed not to be aware that Leśniewski's inquiries were more like an endless quest, not sufficient to become incorporated into the actual practice of science. As for debates around interpretations of Leśniewski's quantifiers, as noted above, Kotarbiński did not mention them in his programme.

Various ways of understanding Leśniewski's ontology (and, specially, its quantifiers), and making it closer to the practice of science, are extensively and insightfully discussed in four chapters of Simons' book (1992) under the attractive title 'Philosophy and Logic in Central Europe from Bolzano to Tarski'.

2.3 A Pragmatically Liberalised Variant of Nominalism

Adherents of nominalism in its reistic version might be particularly interested in Grzegorczyk's book (1997) 'Logic—A Human Affair'. The author declares himself a follower of reism, but reism in a liberal sense. Let the following quotation explains his approach.

> The style or writing I choose in this book may be called reistic. It is reism in a liberal sense. [...] My style of writing presupposes an ontological vision of the world. This ontology focuses on first of all on the reality of everyday experience and describes it in the following way. The reality of every-day experience comprises: things with different properties connected by different relations, making up different sets. This ontology will be called reistic. (Grzegorczyk 1997: 12)

The adjective 'reistic' seems justified by two facts. First, 'things' are listed in the first place, as the basis of describing the world, in contrast to those ontologies which assume a different starting point. For instance, for Whitehead individuals are processes, and for Goodman they are sense-data (cf. Shottenkirk 2009).

Second, the postulate to resort to everyday experience is like Kotarbiński's (1979: 224) claim that 'reists strive to think as common people'. Common people are those who think and talk about the world in terms of everyday sensual experiences concerning things. Moreover, Grzegorczyk may have had in mind that scientists, though highly skilled in most abstract considerations, check their theories against the observable indications of their instruments, and those are things of everyday experience.

However, apart from such reminiscences of the reistic doctrine, Grzegorczyk astonishes us with the following statement,which seems to subvert the very foundations of Leśniewski's and Kotarbiński's nominalism: 'The reistic style of description of the world, in the version presented in this book, may be exhibited in a more technical way as subsumed: either to the simple theory of types or to set theory with basic elements' (Grzegorczyk 1997: 13).

'Basic elements' means individuals, that is, elements of the first type. Then the author refers to the type-theoretical hierarchy of objects founded on basic elements. (1) Basic objects (which are neither sets, nor properties, nor relationships). (2) Properties of basic elements. (3) Relations between basic elements. (4) Properties of the properties of basic elements. (5) Relations between the properties of basic elements and (6) Properties of properties of properties, and ad infinitum.

Sets are considered as determined by properties. Of course, inversely, properties may also be understood as determined by sets, and relations may be considered as properties of pairs. Hence, the whole picture of the world may be formulated in the simple theory of types, as well as in the second of the two technical ways mentioned above, namely Ernst Zermelo's set theory using the so-called 'Uhrelements' (this German term corresponds to the English 'basic elements').

A similar picture of reality is found in Gödel's article 'Russell's Mathematical Logic' (1944). The relevant passage reads as follows.

> By the theory of simple types I mean the doctrine which says that the objects of thought [...] are divided into types, namely: individuals, properties of individuals, relations between individuals, properties of such relations, etc., with a similar hierarchy for extensions [i.e., sets]. (Gödel 1944: 126)

These citations provide an inspiring insight into the core structure of reality; an insight to inquire into in our everyday thinking and scientific proceedings as well. In the universe so constructed there is a place and need for acts of abstraction which start from observably given physical things.

For example, five is a property of the physical structure, like that of my five fingers. We grasp this fact owing to the most basic, that is, the first order level of abstraction. Being a cardinal number is a property of five, and so it belongs to the second order of abstraction. Being an abstract entity is a property of cardinal numbers grasped with acts of still higher abstraction, and so on.

At every level of abstraction, starting from the first, quantifiers are interpreted in the same way, contrary to Kotarbiński's (1957) postulate (cf. 2.2) that quantifiers be conceived differently when applied to individuals and when applied to classes. However, Grzegorczyk's project of liberalised reism does accept Zermelo's set theory which includes, for instance, the axiom of abstraction in which quantifiers of the same kind are applied to the variable x, ranging over individuals, and y that ranges over sets to assert the existence of a set.

$$[\mathrm{AA}]\ \exists y \forall x (x \in y \Leftrightarrow \varphi(x)).$$

Notice that the above formula, when applied without any restrictions, leads to antinomies. We have a restricted secure form, the axiom of subsets, but it is a bit more complicated, while the simplified ('naive') form AA is sufficiently suited to the present argument (cf. Fraenkel 1976: 16).

Grzegorczyk does not use the term 'nominalism', but borrows Kotarbiński's 'reism' to redefine this term in his own original way. Kotarbiński's reism is a variety of nominalism. Is Grzegorczyk's project another case of nominalism, the latter being a more general current? We will try to demonstrate that Grzegorczyk's approach deserves to be called 'moderate nominalism' in contrast to Kotarbiński's 'extreme nominalism'.

The crucial difference lies in the fact that the latter treats abstract expressions as devoid of any cognitive sense, and therefore useless, and demands that they be dispensed with in favour of concrete terms alone. On the other hand, moderate nominalism acknowledges the cognitive usefulness, and even indispensability of abstracts, while the nominalist edge is directed against attributing the same ontological category to both sorts of entities.

According to moderate nominalism, both conditions, that of being indispensable and that of being of a different category, can be satisfied simultaneously. Some indications of how this can be accomplished are present in the simple theory of types as employed by Grzegorczyk. Further insights are provided by Ajdukiewicz's categorial grammar, being a logical and linguistic theory with far-reaching ontological corollaries.

Another version of liberalised (moderate) nominalism, not so formal, but nevertheless thought-provoking, has been suggested by Kotarbińska (1979).

3 Abstraction as the Driving Force of Scientific Progress

3.1 Some Comments on the Relationship Between Rationalism and Nominalism

Though rationalism, that in the platonic tradition, and nominalism, allied with empiricism, have been engaged in a centuries-long controversy, there is a chance for their coming to an agreement—provided that neither of them takes an extreme position.

Extreme rationalism minimises the significance of individuals and of sensory experience, while extreme nominalism exaggerates both. However, pragmatic rationalism appreciates both as the point of departure for the activity of abstraction—the driving force of human cognition. On the other hand, nominalism in its moderately liberal version does not deny the cognitive value or significance of abstract concepts.

This moderate nominalism rightly opposes Platonism in its original form going back to Plato himself. Plato's disinterest in individuals was inherited even by Aristotle, who ignored individual terms in his logic and dealt with general names alone, i.e., those referring to universals. Let us note that in syllogisms all terms must always denote either classes or properties, never individual entities. Hence Aristotle's medieval followers must have treated 'Socrates' like a general name (though restricted to denoting just one entity).

This is why ancient logic, though a marvellous achievement in itself, was far from becoming a driving force of scientific progress. Logic was not able to meet this challenge until Frege initiated symbolic logic in the nineteenth century. Only then was justice done to the role of individual terms in the language of logic and its application in science.

For this reason, and due to an essential difference in appreciating empirical knowledge, historical Platonism cannot be incorporated into pragmatic rationalism. On the other hand, moderate nominalism, like that developed by Grzegorczyk (see 2.3 above), does not require that abstract concepts be abandoned, but only that they be treated with due caution to prevent undesirable ontological commitments.

Further, in this Section, we will discuss the role of Ajdukiewicz's grammar as a theory of syntactic categories (also called syntactic types) in defining ontological types recognised by pragmatic rationalism. Those are conjectured on the basis of the set-theoretical axiom of abstraction whose key function should be duly emphasised. Since this axiom is a typical product of the human mind, unattainable for the senses, this provides us with a convincing justification for the use of the term 'rationalism'.

3.2 Ontological Types as Ones to Be Addressed from a Syntactic Point of View

In Sect. 2.3, we discussed a type-theoretical approach to formal ontology suggested by Grzegorczyk. When calling his view 'reism', using the term introduced by Kotarbiński, Grzegorczyk does not take it literally but (in his own words) in a fairly liberal sense; it appears to be representative of moderate nominalism. We will look at it as a counterbalance to the extreme nominalism of Kotarbiński's camp within the Lvov-Warsaw School.

The discussion culminates with the introduction of the axiom of abstraction (AA). Its other frequently used name is 'axiom of comprehension'. We prefer the former in that its content combines with the epistemological concept of abstraction. The connexion is close, indeed. Moreover, it provides a convenient point of departure for considering ontological types of abstracts.

The issue of ontological types, similar to the issue of ontological categories in Aristotle, is crucial for the controversy between rationalism and nominalism. The key ontological question—aptly stated by Quine—'what is there?'—is answered by listing types of entities: those which exist, and those which do not exist, or rather, those which exist in a somehow weaker way.

The axiom AA is of key ontological importance. It asserts the existence of 'classes' or 'sets'(here these terms are treated as synonyms), and thus implies the next question: which types of entities can be elements of classes? If there are more such types, the next issue arises: does any ordering of types hold? In other words: does any distinguished fundamental type exist on which the other ones are based? If so, which of them performs this basic role?

The inquiry launched by these questions can conveniently start from considering 'syntactic categories' or 'syntactic types', traditionally called 'parts of speech'. As for some types, there is a fairly common agreement among users of ordinary language (though not necessarily among philosophers) concerning the kinds of entities referred to by each type. Namely, nouns denote things, adjectives refer to properties, verbs to actions and activities. As for some other types, as e.g. connectives, it is debatable if they have ontological counterparts, or are only syntactic devices used to build more complex expressions like sentences.

It is up to philosophers to establish which of such common-sense opinions should be accepted, which possibly denied, and how doubtful cases, if any, should be solved. As is well known to students of philosophy, its history is full of heated controversies about each of these points.

To take this challenge, philosophers need a more precise method of analysing languages than that offered by traditional theories of grammar. An excellent toolkit for this purpose is provided by 'categorial grammar' (CG).

3.3 Syntactic Types and Their Relationship to Ontological Types According to CG

In what follows, the term 'category' will be used interchangeably with 'type', as required by convenience and context. In particular, it is convenient to hint in this way at the affinity between categorial grammar and Russell's theory of logical types discussed in the Sect. 2.

CG is an achievement duly attributed to Ajdukiewicz. It was him who played the main role in constructing the theory, though he had quite a few significant predecessors and excellent successors. The former included: Frege, who introduced the category of function to the ontology of logic; Edmund Husserl, who defined the concept of semantic category, present also in Leśniewski's ontology; Jan Łukasiewicz, whose prefix notation paved the way to Ajdukiewicz's algorithm used to decide about syntactic connectivity.

There is also a host of philosophers and logicians much credited for CG, who made some essential contributions to its development and enhancement. They include such eminent authors as Yehoshua Bar-Hillel, Józef M. Bocheński, Max Cresswell, Peter Geach, Henryk Hiż, Joachim Lambek, Richard Montague, Alfred Tarski, or Johan van Benthem. See Buszkowski et al. (1988).

In the core of CG is the following basic idea of the sameness of syntactic categories.

> Two expressions belong to the same syntactic category if and only if a sentence which contains one of these expressions does not cease to be a sentence if one of them is replaced by the other. (Ajdukiewicz 1965)

For instance, the sentence 'Kotarbiński is a nominalist' remains a sentence after we replace 'Kotarbiński' by 'Ajdukiewicz', or '…is a nominalist' by '…is a Platonist'. The former belong to the category of names, the latter—to that of predicates. A much detailed discussion of syntactic categories is found in the book: Buszkowski et al. (1988).

In the set of categories, we single out the subset of two 'basic categories', i.e. category *s* of 'sentences', and *n* of 'names'. The intuition about basic categories is that expressions which belong to them are complete; they correspond to the medieval 'categoremata', as opposed to 'syncategoremata', or 'incomplete' expressions. In the sentence 'Kotarbiński is a nominalist', the name 'Kotarbiński' belongs to the type of complete expressions, while the copula 'is' to incomplete ones. The latter also includes connectives, quantifiers, etc.

Let us look at the semantic status of some syncategorematic expressions. Some instructive examples can be found in the case of logical operators, such as propositional connectives, quantifiers, etc. Consider the following definition of $\Rightarrow$.

Definition 2

Any sentence of the form $p \Rightarrow q$ is true if and only if: either the antecedent is false or the consequent is true.

The symbol $\Rightarrow$ as well as the verb 'implies' belong to the category of syncategorematic words. In order to briefly refer to sentences which satisfy the above definition, we will introduce the symbol '$\Rightarrow$' and the term 'implication'.

Definition 3

'implication' $\stackrel{\text{def}}{=}$ 'sentence of the form $p \Rightarrow q$'.

Now, Definition 2 concerning a syncategorematic word can be replaced by the above definition which refers to implication, hence a logical relationship which is an abstract entity. Thus, 'implication' must belong to the category of abstract entities. As an abstract term, it is defined as follows:

Definition 4

Implication is true if and only if: either its antecedent is false or its consequent is true.

According to the standard of extreme nominalism, abstract terms should be banned as nonsense, unless they are introduced as abbreviations for sensible, comprehensible expressions. This conditions seems to be satisfied in the case of 'implication' as an abbreviation of Definition 2.

This provides us with an example illustrating the nominalist tenet that sentence S*, containing an abstract term, can be admitted to scientific discourse when and only when it is possible to replace S* by a S which is 'synonymous' with S* but which does not contain any abstract terms.

Let us notice, by the way, that there is something mysterious about this condition of synonymity. To be synonymous, two expressions must have the same meaning and reference. Consequently, the sentence containing an abstract term must have a meaning and a reference as well. If so, should we be obligated to get rid of and avoid the one condemned (by nominalists) for dealing with abstracts? If both say the same, they must be cognitively equivalent. It follows that abstract terms must have a reference, hence have a counterpart in the reality, belonging to a relevant ontological type. If we are wrong in this argument, correction by an adherent of extreme nominalism would be most welcome.

Let the same problem be exemplified by yet another illustration: the abstract word 'class', or 'set'. It deserves our special attention as the main target of extreme nominalists' offensive against abstract expressions.

Consider the concept 'a subset of a set' (symbolically represented by the sign of inclusion $\subset$), defined as follows.

Definition 5

Y is a subset of Z $\Leftrightarrow \forall x(x \in \mathrm{Y} \Rightarrow x \in Z)$.

Owing to this definition, in some contexts we can avoid the abstract term 'subset' (unseemly for nominalists) in the *definiendum*, by making use of the *definiens* in Definition 4. Here, it is possible to express one's thought with an abstract-free sentence.

However, there are situations in which abstract terms are indispensable to express to refer to a fact. Consider a sentence as simple as: 'This pair of shoes is mine'. How could one dispense with the name of a set (italicised)? Or (another example) of the name of the relationship 'being greater' (relations are ordered sets) when discussing numbers? This makes the position of extreme nominalism even worse than that discussed above concerning the riddle of equivalence.

3.4 On How the Operator of Abstraction Transforms an Incomplete Expression into a Complete One to Name an Abstract Entity

Let us return to the axiom of abstraction AA in 2.3 above. It asserts that there is a set whose elements satisfy the condition φ. In the simplest case, this condition is expressed by a one-place predicate, say 'P', referring to a property. In some more involved cases, it may be a more-place predicate (to refer to a relation), or a compound sentence in the form of negation, conjunction, or anything else of the kind.

Consider the formula which asserts the existence of a set y whose elements are those and only those individuals who have the property *R*. Let us employ a special notation to distinguish various orders of types. A variable ranging over individuals belongs to order one, that ranging over classes of individuals to order two, while in the case of a set of classes of individuals we have order three, and so on—ad infinitum.

When equipped with this notation, the axiom of abstraction at the lowest level may look as follows:

$$[\mathrm{AA}*]\ \exists y^2 \forall x^1 (x^1 \in y^2 \Leftrightarrow R(x^1)).$$

Now, on the basis of AA* we can create names of abstract entities whose existence is granted by the occurrence of '$\exists y^2$'. This creation is accomplished by a logical operator, the so-called 'abstraction operator' typographically formed by a colon preceded by the given variable, and followed by the formula in question, the whole enclosed by curly brackets. In the set-theoretical notation, this symbolises the set whose elements are listed between brackets (e.g., {1, 2, 3} is a set of the first three natural numbers after 0).

Let '*R*' be read as 'is a rationalist'. Then the class of individuals who are rationalists is referred to by the following abstract name:

$$y^2 = \left\{x^1 : \mathrm{R}\left(x^1\right)\right\}.$$

In this way, an incomplete expression, i.e. the predicate *R*—'is a rationalist'—is transformed into a complete one, that is, the name of a class containing rationalists.

Such a procedure can be continued indefinitely. By using ever greater numerals to denote ever greater orders of types, we obtain more and more abstract concepts. This gives rise to an interesting elucidation of both the rationalist and the nominalist stance. To this end, we should take a moment to consider the structure of reality.

3.5 On Abstract Constituents of Physical Reality

In Hao Wang's 'Reflections on Kurt Gödel' (1990) and other records of conversations with Kurt Gödel, e.g. Wang's (1997), there is a very interesting remark which Gödel made on 'fiveness' as an example of numbers as properties of things, located in physical reality. The paragraph which follows is a compilation of several statements made by Gödel as recorded by Wang (and fully endorsed by the author of this paper).

> Take small and simple physical sets, such as the set of five petals on the flower in front of me, or five fingers on my left hand. Their perception seems analogous to the perception of simple physical objects and their properties, e.g. the redness of a flower. Numbers are embodied physically, they are shapes that repeat in natural processes. 'Fiveness' exists in a plant's genetic code before and after its physical embodiment in petals. (Wang 1997: 202)

We will need a suitable term to refer to the category represented by 'fiveness' in Gödel's statements. Let us look for other instances to find a useful generalisation.

When treating individuals as fundamental and autonomous 'constituents of reality', rationalism acknowledges the existence of other kinds of constituents, not so autonomous in their existence, but just as real. We call them 'abstract constituents' in the etymological Latin sense of *abstractio* meaning 'separation'.

Geometry is a domain in which we can find nice examples of constituents of a physical spatiotemporal whole which can be mentally separated from their wholes. The fact that something cannot be physically separated does not lessen its being as real as the whole in question.

For instance, the moon participates in reality as an independent entity. Its shape, surface, diameter etc. participate in it as well, but as dependent parts of the moon which cannot be physically separated from it. We can separate them only mentally. However, this does not cancel their being real as indispensable constituents of the moon. It is impossible for a surface to exist without a solid, but it is equally impossible for a solid to exist without its surface. The same should be said about planes whose abstract constituents are lines, and about points being abstract constituents of lines.

Were the degrees of reality recognised by senses, then the surface of a solid might appear more real than the solid itself. For, like with the moon, what we see is only the surface, and the fact it is actually a solid is a matter of guesswork.

We have a similar situation in arithmetics. The 'fiveness' of my fingers is not an independent thing, but is an abstract constituent of a solid, that is, my five fingers considered as a physical whole.

3.6 The Issue of Correspondence Between Syntactic and Ontological Types: Its Philosophical Import

In the contention between rationalism and nominalism, there are two issues requiring our careful attention. One is that raised by Gödel in his remarks quoted above.

The other is about mapping the set of syntactic types onto ontological types. As for the list of syntactic types, it is relatively easy to reach an agreement about its content. The controversy is about mapping the set of syntactic types onto the ontological ones. Does every item on the former list has a counterpart on the latter? According to extreme nominalism, only individual names have ontological counterparts, and general names if predicated about individuals.

As for the full-fledged rationalistic approach, a good example is found in Suszko's contribution in the 'Festschrift' (1964) offered to Ajdukiewicz. The paper deals with syntactic categories and their denotations (i.e., references) in formalised languages. This mapping is perfectly one-to-one. Individual terms, e.g. '0' refer to individuals, while the rest to various types of functions, differentiated with type indexes. These are taken from Ajdukiewicz's notation which is designed to perform algorithmic testing of syntactic correctness. In this study, functions of arbitrarily high order are admitted.

Such a mapping is available in formalised languages. However, in ordinary languages, used also in empirical sciences, this is not the case. As noticed by Ajdukiewicz (1960: 209), there are ontological types which can be referred to by expressions of various syntactic types. For instance, the state of being red is referred to by the predicate '…is red', the abstract property-name 'redness', and the abstract class-name '{*x*: *x* is red}'.

For pragmatic rationalism, the problem of mapping syntactic types is of lesser significance. In practice, when using any expression to speak of the world, one believes that they are not speaking about nothing; hence not only nouns, but also adjectives, verbs etc. must have references. In such contexts, philosophical beliefs are, in fact, irrelevant to the practical use of language, and to all human activities entailing the use of language.

However, this is not irrelevant for extreme nominalists. They are committed to philosophical correctness and purity, hence they advise people to avoid abstract expressions, wherever possible, by replacing them with concrete terms. When they are not available, it is so because of the imperfection of the existing language, and ought to be remedied by philosophers as soon as possible.

4 Conclusion: On the Progress of Science Owing to Ever Higher Levels of Abstraction

In a book on algebraic topology, William S. Massey makes the following comment:

> The history of mathematics in the last two hundred years or so has been characterised by the considerations of mathematical systems on ever higher levels of abstraction. Presumably this trend will continue in the future. (Massey 1991: 84)

This observation is concerned with the historically most recent stage of mathematics, while Gödel's reflections (those in the box above) refer to its most primitive stage, the point of departure—something like counting on five fingers. This leads us to an overall framework of the main current in the evolution of science: the stream which consists in the progress of increasingly abstract mathematics.

This picture of evolution appears to be in perfect agreement with the rationalist philosophy of science, while at the same time being the opposite of the extreme nominalist programme which says that science should get rid of abstract concepts. However, a balanced brand of nominalism (in principle understanding the role of abstraction) puts forward some arguments about the need for caution when dealing with abstract concepts.

These arguments are connected with a feature of mathematics which unexpectedly appeared by the end of the nineteenth century in the form of set-theoretical antinomies. This experience subverted the centuries-long belief of mathematicians, as well as philosophers, that mathematics was an infallible and perfectly certain discipline, not endangered by any inconsistencies.

The danger in mathematics stemmed from its having been founded on the conceptual apparatus of set theory where some excessive abstractions (like the concept of the set of all sets) proved responsible for antinomies. Another concern has been caused by Gödel's (1931) theorem that it is not possible to prove the consistency of the number theory using procedures available in that theory. More efficient means might have been taken from the set theory, but that theory turned out to be a weak spot in the edifice of mathematics.

This concern about the consistency of mathematics (a situation never faced by earlier mathematicians) invited suggestions as to remedies, and one of them was due to some well-balanced nominalism, like that of David Hilbert. His suggestions, though highly sophisticated, did not succeed, as demonstrated by Gödel (1931); however the tendency towards greater caution proved its validity.

In the case of antinomies, the first remedy consists in detecting and removing the source of inconsistency. This has been done to cure the current set theory, one of the measures having been Russell's theory of types, another one—procedures of axiomatization to result, e.g., in ZF.

However, such provisional remedies cannot ensure that other inconsistencies will not appear, especially when we continue to develop our theory using new highly abstract concepts. A certain measure, which seems to be highly relevant, has been suggested by Gödel and other pragmatist followers of rationalism, like Quine and Hilary Putnam.

It consists in treating mathematics on equal footing with such advanced empirical sciences as physics. This means, obviously, losing the good old privilege of infallibility. However, in the sequence of degrees of fallibility, mathematics and logic are found at the least degree, that is, they enjoy the position of being closest to infallibility.

The philosopher who contributed the most to realising the fallibility of sciences, and the ensuing necessity of testing theories through their application, was the founding father of pragmatism—Charles Sanders Peirce. His pragmatism was closely tied with epistemological rationalism, hence the relevance of the concept of 'pragmatic rationalism'.

Working along these lines, Quine (1953: 42–46) outlined the structure of the entire field of sciences as located at various degrees of infallibility. Gödel's significant gloss to this scenario is to the effect that within the field of mathematics itself there are degrees of certainty as well. For instance, arithmetics enjoys it to a greater degree than some high regions of set theory. In fact, even in arithmetics there is, in this respect, some ordering: the closest to infallibility is the part of arithmetics which deals with small numbers, that is, those used most frequently in human practice such as accountancy, engineering, and so on. Several thousand years of operations performed on such numbers without encountering any inconsistencies provide us with as great empirical evidence of consistency as is humanly possible.

Let us now return to Gödel's story of 'fiveness' as told above (cf. 3.5). It starts from the phrase 'the small physical sets…'. Our perception of such structures is the closest to certainty on account of those very attributes: of being physical and being small (hence entirely within our field of sensory perception). When considering Gödel's two examples, five petals and five fingers, it may prove instructive to combine them with the idea that a fertile abstraction may be due to graphical operations like drawing or writing.

Let us imagine an anonymous genius thousands of years ago who conceives the idea of a graphic representation of those five-element classes. He uses a stick to draw five very small horizontal lines in the sand, followed by the drawing of five petals, which is to mean: as many petals as there are lines. Then he performs a similar operation concerning five fingers. He may produce quite a number of such drawings, repeating the same sequence of lines and completing each with ever new pictures of an observable entity having a similar five-element structure (say, a group of five horses, etc.).

In the next phase of his performance, he erases all the pictures he had drawn. This physical act represents an act of abstraction. Now all the pictures become irrelevant, and what occupies his whole attention is just the identity of all these drawings—each consisting of as many lines as every other one. When remembering later these five sequences of lines, one can use it as a symbol of all such structures. If he prefers a more convenient notation, he may replace the lines with some other conventionally chosen sign, e.g. the Roman 'V'. Thus arise 'digits' whose 'references' produce physical structures of lines. And these are what we call 'numbers'.

This tale makes us aware of some rational core of the nominalist contention. In fact, in order to do science we need from the very beginning some physically observable entities forming discrete structures, and this is what nominalists of the reistic brand call 'things'. If the actual world did not consist of discrete things, but was made of some undifferentiated matter, e.g. sand or water, then presumably no idea of digits or natural numbers would ever occur to us.

Such a story of the origin of various kinds of numbers might be continued in several chapters, each concerned with a new and more extensive class of numbers. Its being more extensive will be due to abstracting from some features of the former class. For instance, an experience with dividing some physical entity into parts would lead to the idea of fractions. Next we add the class of fractions to the class of natural numbers, and so obtain a more extensive class; when defining it, we abstract from differences between the constituent subclasses, and focus on what is common to them. Such abstraction yields the set of rational numbers.

Other experiments, also based on physical substrates, would result in the notion of negative numbers, then in the notion of reals, etc. When having such a physical basis, a mathematical theory might be tested in a way resembling (though not identical with) the testing of physical theories. This feature scores a point for moderate nominalism, that is, one which does not deny the cognitive indispensability of higher levels of abstraction.

Let us conclude. Pragmatic rationalism ought to be appreciated both for doing justice to the empirical origins of science, and to the power of abstraction as the driving force of scientific progress. On the other hand, pragmatic nominalism, at least in a form similar to Grzegorczyk's liberal reism, has the merit of emphasising the role of ordinary language and everyday experience. Thus, a friendly collaboration between these orientations could open a new promising chapter in the splendid history of the Lvov-Warsaw School.

References

Ajdukiewicz, K. (1960). *Język i poznanie. Wybór pism z lat 1920–1939* [A selection of writings 1920–1939], vol. I. Warszawa: PWN.

Ajdukiewicz, K. (1965). *The foundations of statements and decisions*. Warszawa: PWN.

Bacon, F. (2011). The new organon. In J. Spedding, R. L. Ellis, & D. D. Heath (Eds.), *The works of Francis Bacon* (Vol. 4, pp. 39–248). Cambridge: Cambridge University Press.

Buszkowski, W., Marciszewski, W., van Benthem, J. (Eds,). (1988). *Categorial grammar*. Amsterdam: Benjamins.

Colyvan, M. (1998). Defence of indispensability. *Philosophia Mathematica, 6*(1), 39–62.

Fraenkel, A. (1976). *Abstract set theory*. Amsterdam: North Holland Publishing Company.

Gödel, K. (1931). Über formal unentscheidbare Sätze der PrincipiaMathematica und verwandter Systeme I. Monatshefte für Mathematik und Physik, *38*, 173–198.

Gödel, K. (1944). Russell's mathematical logic. In S. Feferman, J. Dawson, & S. Kleene (Eds.), *Journal of symbolic logic* (pp. 119–141). Evanston, IL: Northwestern University Press.

Grzegorczyk, A. (1997). *Logic—A human affair*. Warszawa: Wyd. Naukowe Scholar.

Kotarbińska, J. (1979). Puzzles of existence. In J. Pelc (Ed.), *Semiotics in Poland, 1894–1969*. Warszawa and Dordrecht: Kluwer, PWN.

Kotarbiński, T. (1957). *Wykłady z dziejów logiki* [Lectures on the history of logic]. Łódź: Ossolineum.

Kotarbiński, T. (1979). The reistic or concretistic approach. In J. Pelc (Ed.), *Semiotics in Poland* (pp. 1894–1969). Warszawa and Dordrecht: Kluwer, PWN.

Küng, G. (1981). Leśniewski's systems. In W. Marciszewski (Ed.), *Dictionary of logic as applied in the study of language: Concepts, methods, theories* (pp. 168–177). Hague: Nijhoff.

Marciszewski, W. (1988). A chronicle of categorial grammar. In W. Buszkowski, W. Marciszewski, & J. van Benthem (Eds.), *Categorial grammar* (pp. 7–21). Amsterdam: Benjamins.

Marciszewski, W. (2012). Confrontation of reism with type-theoretical approach and everyday experience. *Studies in Logic, Grammar and Rhetoric, 27*(40), 107–125.

Marciszewski, W. (2017). Scientific philosophy in the Lvov-Warsaw school: Pragmatic rationalism as its main trend. In P. Polak et al. (Eds.), *Oblicza filozofii w nauce* [*Faces of philosophy in science: A Festschrift in honour of Michał Heller*]. Kraków: Copernicus Center Press.

Massey, W. S. (1991). *A basic course in algebraic topology*. Berlin and Heidelberg: Springer Science & Business Media.

Nelson, M. (2012). Existence. *Stanford encyclopedia of philosophy*. https://plato.stanford.edu/entries/existence.

Quine, W. V. O. (1953). *From a logical point of view*. Cambridge, MA: Harvard University Press.

Shottenkirk, D. (2009). *Nominalism and its aftermath: The philosophy of Nelson Goodman*. Berlin and Heidelberg: Springer Science & Business Media.

Simons, P. (1992). *Philosophy and logic in Central Europe. From Bolzano to Tarski*. Dordrecht: Kluwer Academic Publishers.

Suszko, R. (1964). O kategoriach syntaktycznych i denotacjach wyrażeń w językach sformalizowanych [On syntactic categories and references of expressions in formalised languages]. In *Rozprawy logiczne. Księga pamiątkowa ku czci profesora Kazimierza Ajdukiewicza* [Logical Dissertations. Festschrift in Honour of Professor Kazimierz Ajdukiewicz]. Warszawa: PWN.

Wang, H. (1990). *Reflections on Kurt Gödel*. Cambridge, MA: MIT Press.

Wang, H. (1997). *A logical journey: From Gödel to philosophy*. Cambridge, MA: MIT Press.

11

Some Problems Concerning Axiom Systems for Finitely Many-Valued Propositional Logics

Mateusz M. Radzki

1 Introduction

In this chapter, we will examine some problems concerning axiom systems for finitely many-valued propositional logics. It has been recently demonstrated that neither the known axiom system for the functionally complete three-valued logic nor the particular method of constructing axiom systems for finitely many-valued logics satisfies some salient metalogical requirements.

Firstly, we will examine an axiom system for the functionally complete three-valued logic based on the well-known axiom system for the three-valued logic developed by Jan Łukasiewicz. The examined axiom system was introduced by Jerzy Słupecki. As we will see, this axiom system is not semantically complete.

M. M. Radzki (✉)
Institute of Philosophy and Sociology,
The Maria Grzegorzewska University, Warsaw, Poland
e-mail: matradzki@wp.pl

A. Drabarek et al. (eds.), *Interdisciplinary Investigations into the Lvov-Warsaw School*, History of Analytic Philosophy,
https://doi.org/10.1007/978-3-030-24486-6_11

Then, we will consider a method of constructing axiom systems for standard many-valued propositional logics introduced by John B. Rosser and Atwell R. Turquette. We will demonstrate that the Rosser-Turquette method fails to produce adequate axiom systems for a particular class of many-valued propositional logics, i.e., for the many-valued propositional logics developed by Łukasiewicz.

2 Many-Valued Logics of Łukasiewicz and the Functionally Complete Three-Valued Logic

In the works published in the early 1920s, the Polish logician and philosopher Jan Łukasiewicz introduced the third truth-value ½, intermediate between 1 (interpreted as 'true') and 0 (interpreted as 'false'). This way, Łukasiewicz initiated philosophical and mathematical investigations on many-valued logics, first—on finite calculi, and then—on infinite ones (see, for example, Malinowski 2006: 549; Woleński 1989: 124–125).

As is well-known, his discovery was a result of philosophical debates within the Lvov-Warsaw School concerning, for example, the problems of induction and the theory of probability, the question of determinism, indeterminism and related problems concerning causality and modality (Malinowski 2009: 81–82; Radzki 2017a: 404; Woleński 1989: 124).

In the three-valued propositional logic developed by Łukasiewicz (hereinafter $Ł_3$), the propositional connectives $\sim, \rightarrow, \vee, \wedge, \leftrightarrow$ are characterised by the following truth-tables (Łukasiewicz 1967: 54; see also Chen and Pham 2001: 65; Malinowski 2006: 546; Nguyen and Walker 2000: 66; Tooley 1997: 134; Wan and Chen 2014: 196) (Table 1).

According to Mordchaj Wajsberg, taking the rule of substitution and the rule of detachment, $Ł_3$ is described by the following four axioms (Wajsberg 1967: 264; see also Bergmann 2008: 100; Cignoli et al. 2000: 87–88; Ciucci and Dubois 2012: 155; Goldberg and Weaver 1982: 240; Gottwald 2001: 195; Kabziński 1981: 206; Słupecki et al. 1967: 52):

Table 1 The definitions of the propositional connectives $\sim, \rightarrow, \vee, \wedge, \leftrightarrow$ in the three-valued logic of Łukasiewicz (In the truth-tables for binary propositional connectives, the truth-value of the first argument is given in the vertical line, the truth-value of the second argument is given in the horizontal line, and the outcome is given in the intersection of these lines)

p	$\sim p$	$\rightarrow$	0	½	1	$\vee$	0	½	1	$\wedge$	0	½	1	$\leftrightarrow$	0	½	1
0	1	0	1	1	1	0	0	½	1	0	0	0	0	0	1	½	0
½	½	½	½	1	1	½	½	½	1	½	0	½	½	½	½	1	½
1	0	1	0	½	1	1	1	1	1	1	0	½	1	1	0	½	1

W1. $p \rightarrow (q \rightarrow p)$;
W2. $(p \rightarrow q) \rightarrow ((q \rightarrow r) \rightarrow (p \rightarrow r))$;
W3. $(\sim p \rightarrow \sim q) \rightarrow (q \rightarrow p)$;
W4. $((p \rightarrow \sim p) \rightarrow p) \rightarrow p$.

We can introduce the remaining propositional connectives into the system using the following definitions:

Definition 1

$$(p \vee q) \overset{\text{DEF}}{\leftrightarrow} ((p \rightarrow q) \rightarrow q);$$
$$(p \wedge q) \overset{\text{DEF}}{\leftrightarrow} \sim (\sim p \vee \sim q);$$
$$(p \leftrightarrow q) \overset{\text{DEF}}{\leftrightarrow} ((p \rightarrow q) \wedge (q \rightarrow p)).$$

In a 1936 article (Słupecki 1967: 335–337; see also Bolc and Borowik 1992: 64), in order to create a functionally complete three-valued logic, Słupecki extended the axiom system W1–W4 with another two axioms:

W5. $Tp \rightarrow \sim Tp$;
W6. $\sim Tp \rightarrow Tp$.

Słupecki's T-function is defined by the following truth-table (see, for example, Date 2007: 138; Epstein 1993: 323) (Table 2).

However, it has been recently proved that Słupecki's axiom system is not semantically complete (Radzki 2017a: 403–415).

Table 2 Słupecki's *T*-function

p	Tp
0	½
½	½
1	½

Let us examine the following tautologies of the functionally complete three-valued logic:

$$Tp \rightarrow \sim Tq;$$
$$\sim Tq \rightarrow Tp.$$

Thus, if Słupecki's axiom system is semantically complete, the above formulae are provable in W1–W6.

However, the opposite result can be obtained as well. The proof follows a proof-theoretic method of constructing independence proofs, and it leads to the conclusion that $Tp \rightarrow \sim Tq$ and $\sim Tq \rightarrow Tp$ are not provable in W1–W6 (Radzki 2017a: 411).

3 The Rosser-Turquette Method

Let us now examine a certain general method of constructing axiom systems. Such a method for standard many-valued propositional logics was introduced by Rosser and Turquette (1952).

The notion of 'standard' logics involves the notion of 'standard' conditions of the propositional connectives and the truth-functions associated with them (Rosser and Turquette 1952: 25). It expresses 'classical' properties of the propositional connectives, and thus allows us to simplify some problems of the metatheory of many-valued logic.

Now, assume $n \geq 2$ ($n \in \mathbb{N}$) and $1 \leq k < n$. The set of all truth-values, i.e., E_n, is such that:

$$\mathrm{E}_n = \{1, 2, \ldots, n\},$$

where 1 stands for 'true', and n stands for 'false'.

The set of designated truth-values, i.e., D_k, is as follows:

$$\mathrm{D}_k = \{1, 2, \ldots, k\}.$$

The propositional language of the many-valued logics concerned is given by the matrix

$$\mathfrak{M}_{n,k} = (\mathrm{U}_n, \mathrm{D}_k),$$

where $\mathrm{U}_n = (\mathrm{E}_n, f_1, \ldots, f_m)$ and $f_1, \ldots, f_m$ are the truth-functions.

Let us examine some propositional connectives, i.e., negation ($\neg$), implication ($\Rightarrow$), disjunction ($\vee$), conjunction ($\wedge$), equivalence ($\Leftrightarrow$) and special unary connectives $j_1, \ldots, j_n$. The connectives $\neg$, $\Rightarrow$, $\vee$, $\wedge$ and $j_1, \ldots, j_n$ satisfy standard conditions, if for any $x, y \in \mathrm{E}_n$ and $i \in \{1, 2, \ldots, n\}$ the following definitions hold (Rosser and Turquette 1952: 25–26; see also Malinowski 2006: 552):

Definition 2

$$\neg x \in \mathrm{D}_k \text{ iff } x \notin \mathrm{D}_k,$$
$$x \Rightarrow y \notin \mathrm{D}_k \text{ iff } x \in \mathrm{D}_k \text{ and } y \notin \mathrm{D}_k,$$
$$x \vee y \in \mathrm{D}_k \text{ iff } x \in \mathrm{D}_k \text{ or } y \in \mathrm{D}_k,$$
$$x \wedge y \in \mathrm{D}_k \text{ iff } x \in \mathrm{D}_k \text{ and } y \in \mathrm{D}_k,$$
$$x \Leftrightarrow y \in \mathrm{D}_k \text{ iff either } x, y \in \mathrm{D}_k \text{ or } x, y \notin \mathrm{D}_k,$$
$$j_i(x) \in \mathrm{D}_k \text{ iff } x = i.$$

Now, we can introduce the definition of the notion of standard logics.

Definition 3
Every many-valued propositional logic that contains standard connectives as primitive or definable is a standard one (Rosser and Turquette 1952: 25–26; see also Malinowski 2006: 552).

In line with Rosser and Turquette, taking the rule of detachment, every standard many-valued propositional logic is axiomatizable by means of the set of standard connectives, i.e., $\mathrm{F} = \{\Rightarrow, j_1, \ldots, j_n\}$ and the following seven axiom schemata (Malinowski 2007: 42):

(A1) $A \Rightarrow (B \Rightarrow A)$,
(A2) $(A \Rightarrow (B \Rightarrow C)) \Rightarrow (B \Rightarrow (A \Rightarrow C))$,
(A3) $(A \Rightarrow B) \Rightarrow ((B \Rightarrow C) \Rightarrow (A \Rightarrow C))$,
(A4) $(j_i(A) \Rightarrow (j_i(A) \Rightarrow B)) \Rightarrow (j_i(A) \Rightarrow B)$,
(A5) $(j_n(A) \Rightarrow B) \Rightarrow ((j_{n-1}(A) \Rightarrow B)$
$\Rightarrow (\ldots \Rightarrow ((j_1(A) \Rightarrow B) \Rightarrow B)\ldots))$,
(A6) $j_i(A) \Rightarrow A$ (for $i = 1, 2, \ldots, k$),
(A7) $j_{i(r)}(A_r) \Rightarrow \big(j_{i(r-1)}(A_{r-1})$
$\Rightarrow \big(\ldots \Rightarrow \big(j_{i(1)}(A_1) \Rightarrow j_v(\mathcal{F}(A_1, \ldots, A_r))\big)\ldots\big)\big)$

(for $i = 1, 2, \ldots, n$, for each propositional connective $\mathcal{F} \in$ F, and for the particular truth-value $v = f(i(1), \ldots, i(r))$, where f is a truth-function corresponding to $\mathcal{F}$) (compare with Gottwald 2001: 109).

In the literature, the Rosser-Turquette method is usually described as the formal tool that tackles the problem of axiomatizability of a particular class of many-valued logics, i.e., standard many-valued propositional logics, including finitely many-valued propositional logics of Łukasiewicz (hereinafter $Ł_n$) and Emil Post (cf. Malinowski 2006: 553).

Then, let us consider $Ł_n$. Every $Ł_n$ is given by the matrix:

$$\mathfrak{M}_{Ł_n} = \left(\mathrm{E}_{Ł_n}, \sim, \rightarrow, \wedge, \vee, \leftrightarrow, \{1\}\right),$$

where $\mathrm{E}_{Ł_n} = \{0, 1/(n-1), 2/(n-1), \ldots, (n-2)/(n-1), 1\}$. 1 is a designated truth-value, and stands for 'true'; 0 stands for 'false'. The propositional connectives $\sim$ and $\rightarrow$ are defined by the following arithmetical rules (Malinowski 2007: 42):

Definition 4

$$\sim x = 1 - x,$$
$$x \rightarrow y = \min\,(1, 1 - x + y),$$

The remaining connectives are characterised as follows:

Definition 5

$$x \vee y = (x \to y) \to y = \max(x, y),$$
$$x \wedge y = \sim(\sim x \vee \sim y) = \min(x, y),$$
$$x \leftrightarrow y = (x \to y) \wedge (y \to x) = 1 - |x - y|.$$

Since every $Ł_n$ is a standard logic, $\Rightarrow$, $j_1, \ldots, j_n$ are definable by means of $\sim$ and $\to$ (Malinowski 2006: 553; see also Malinowski 2007: 41–42). However, if the Rosser-Turquette axiom system for any $Ł_n$ is supposed to be semantically complete, then $\sim$ and $\to$ (and thus, all remaining connectives of $Ł_n$) have to be definable by means of $\Rightarrow$, $j_1, \ldots, j_n$.

However, it has been recently proved that the mutual definability of the Rosser-Turquette standard connectives and the connectives of $Ł_n$ cannot be achieved simultaneously (Radzki 2017b: 27–32). If $\Rightarrow$, $j_1, \ldots, j_n$ are supposed to be definable in terms of $\sim$ and $\to$, then $\Rightarrow$, $j_1, \ldots, j_n$ have to be normal ones. But it has been proved that $\to$ is indefinable by means of normal $\Rightarrow$, $j_1, \ldots, j_n$. Thus, every Rosser-Turquette axiom system for $Ł_n$ that satisfies the necessary condition of being a sound system is semantically incomplete. Therefore, the Rosser-Turquette method fails to produce adequate axiom systems for $Ł_n$.

4 Conclusion

The above argument reveals the importance of the issue of definability with regard to propositional connectives in metalogical investigations. Since the Rosser-Turquette standard connectives are not equal to the basic connectives of Łukasiewicz's many-valued logics, i.e., $\sim, \to, \vee, \wedge, \leftrightarrow$, the Rosser-Turquette axiom schemata are not expressed in the propositional language of Łukasiewicz's many-valued logics. Investigations concerning the propositional language of the Rosser-Turquette axiom schemata lead to the conclusion that Rosser-Turquette cannot produce adequate axiom systems for $Ł_n$.

It is worth noting that the definability of propositional connectives in terms of the basic connectives of Łukasiewicz's many-valued logics has been described in literature. Since the connectives $\vee, \wedge, \leftrightarrow$ are definable in terms of $\{\sim, \rightarrow\}$, it is sufficient to consider $\sim$ and $\rightarrow$. The following theorem, established by Robert McNaughton (1951: 12), provides an answer to the question whether or not a certain propositional connective is definable in terms of $\{\sim, \rightarrow\}$:

Theorem

Let L_m be a finite-valued propositional logic of Łukasiewicz, i.e., $m > 1$, and let K_m be a finite set of truth-values in a particular L_m, i.e.,

$$K_m = \left\{0, \frac{1}{m-1}, \frac{2}{m-1}, \ldots, 1\right\}.$$

Then, for all $n > 0$, $m > 1$ and for every function f from K_m^n into K_m, f defined for each $x_1, \ldots, x_n$ where every $x_i = s_i/(m-1)$ for some integer s_i, $0 \leq s_i \leq m-1$, is defined in terms of $\{\sim, \rightarrow\}$ if and only if f satisfies the following two conditions

1. f is single-valued, and for $x_i \in K$, $i = 1, \ldots, n$, $f(x_1, \ldots, x_n) \in K$.
2. For each $\left(\frac{s_1}{m-1}, \ldots, \frac{s_n}{m-1}\right)$, if d is the greatest common divisor of

$$s_1, \ldots, s_n, (m-1),$$

then d divides s, where

$$\frac{s}{m-1} = f\left(\frac{s_1}{m-1}, \ldots, \frac{s_n}{m-1}\right).$$

Although by means of the McNaughton theorem we are capable of answering the question whether or not a certain propositional connective is definable in terms of $\{\sim, \rightarrow\}$, we do not have a metalogical method that would allow us to answer the question whether or not either $\sim$ or $\rightarrow$ is definable by means of a particular class of propositional connectives, for example, the class of standard connectives.

To summarise, although, as Słupecki demonstrated in his article (Słupecki 1967: 335–337; see also Evans and Schwartz 1958: 267–270; Karpenko 1989: 465–478; Rose 1981: 113–129; Skala 1988: 1–9; Woleński and Zygmunt 1989: 401–411), every propositional connective of a functionally complete three-valued logic is definable in terms of $\{\sim, \rightarrow, T\}$, Słupecki's axiom system W1–W6 is semantically incomplete. In other words, Słupecki's two axioms added to Wajsberg's axiom system are not sufficient to obtain all tautologies of a functionally complete three valued logic which are constructed out of $\sim$, $\rightarrow$ and T. This conclusion is contrary to the commonly held view that a functionally complete three-valued logic is axiomatised by W1–W6 (see Bergmann 2008: 94; Bolc and Borowik 1992: 64; Mancosu 2010: 97; Mancosu et al. 2009: 318–470; Urquhart 2001: 249–295).

Moreover, although a number of general methods of constructing axiom systems for many-valued logics are described in literature, for example the method of Revaz Grigolia for n-valued ($n > 3$) logics of Łukasiewicz (Grigolia 1997: 81–92), the systems of Marek Tokarz (1974: 21–24) and Roman Tuziak for n-valued ($n \geq 2$) logics of Łukasiewicz (Tuziak 1988: 50), and the method of Słupecki for functionally complete n-valued ($n > 3$) logics (Słupecki 1939: 110–128; see also Bolc and Borowik 1992: 79–82), the only known axiom system for a functionally complete three-valued calculus with a single designated truth-value is that constructed by Słupecki. Therefore, our result indicates the need for constructing a new axiom system for a functionally complete three-valued propositional logic.

Finally, we have demonstrated that the Rosser-Turquette method fails to produce adequate axiom systems for $Ł_n$. This conclusion reveals the importance of the issue of definability with regard to propositional connectives in constructing a metatheory of many-valued logics. Consequently, the argumentation presented above discloses some significant problems with regard to axiom systems for finitely many-valued propositional logics. And last but not least, it is worth noting that the examined issues refer to the philosophical core of many-valued calculi and one of the most important logical achievements of the Lvov-Warsaw School—the many-valued logics developed by Łukasiewicz.

References

Bergmann, M. (2008). *An introduction to many-valued and fuzzy logic*. Cambridge: Cambridge University Press.

Bolc, L., & Borowik, P. (1992). *Many-valued logics 1: Theoretical foundations*. Berlin and Heidelberg: Springer.

Chen, G., & Pham, T. T. (2001). *Introduction to fuzzy sets, fuzzy logic, and fuzzy control systems*. London, New York: CRC Press.

Cignoli, R. L. O., D'Ottaviano, I. M. L., & Mundici, D. (2000). *Algebraic foundations of many-valued reasoning*. Dordrecht: Springer.

Ciucci, D., & Dubois, D. (2012). Three-valued logics for incomplete information and epistemic logic. In R. Goebel, et al. (Eds.), *Lectures notes in artificial intelligence* (pp. 147–159). Berlin and Heidelberg: Springer.

Date, C. J. (2007). *Logic and databases: The roots of relational theory*. Victoria, BC: Trafford Publishing.

Epstein, R. L. (1993). *Multiple-valued logic design: An introduction*. Bristol, Philadelphia: Institute of Physics Publishing.

Evans, T., & Schwartz, P. B. (1958). On Słupecki's T-functions. *Journal of Symbolic Logic, 23*, 267–270.

Goldberg, H., & Weaver, G. (1982). A strong completeness theorem for three-valued logic: Part I. In H. Leblanc (Ed.), *Existence, truth, and provability* (pp. 240–246). Albany: State University of New York Press.

Gottwald, S. (2001). *A treatise on many-valued logics*. Baldock, UK: Research Studio Press.

Grigolia, R. (1997). Algebraic analysis of Łukasiewicz-Tarski's *n*-valued logical systems. In E. Wójcicki & G. Malinowski (Eds.), *Selected papers on Łukasiewicz sentential calculi* (pp. 81–92). Wrocław: Ossolineum.

Kabziński, J. K. (1981). Many-valued logic. In R. Marciszewski (Ed.), *Dictionary of logic as applied in the study of language* (pp. 201–209). Netherlands: Springer.

Karpenko, A. S. (1989). Characterization of prime numbers in Łukasiewicz's logical matrix. *Studia Logica, 48*, 465–478.

Łukasiewicz, J. (1967). Philosophical remarks on the many-valued systems of propositional logic. In S. McCall (Ed.), *Polish logic 1920–1939* (pp. 51–77). Oxford: Clarendon Press.

Malinowski, G. (2006). Many-valued logic. In D. Jacquette (Ed.), *A companion to philosophical logic* (pp. 545–561). Oxford: Blackwell Publishing.

Malinowski, G. (2007). Many-valued logic and its philosophy. In D. M. Gabbay & J. Woods (Eds.), *Handbook of the history of logic—Volume 8: The many valued and nonmonotonic turn in logic* (pp. 13–94). Amsterdam: North-Holland.

Malinowski, G. (2009). A philosophy of many-valued logic: The third logical value and beyond. In S. Lapointe, et al. (Eds.), *The golden age of Polish philosophy: Logic, epistemology and the unity of science* (Vol. 16, pp. 81–92). Berlin: Springer.

Mancosu, P. (2010). *The adventure of reason: Interplay between philosophy mathematics and mathematical logic 1900–1940*. Oxford: Oxford University Press.

Mancosu, P., Zach, R., & Badesa, C. (2009). The development of mathematical logic from Russell to Tarski. In L. Haaparanta (Ed.), *The development of modern logic* (pp. 318–470). New York and Oxford: Oxford University Press.

McNaughton, R. (1951). A theorem about infinite-valued sentential logic. *Journal of Symbolic Logic, 16*(1), 1–13.

Nguyen, H. T., & Walker, E. A. (2000). *A first course in fuzzy logic*. London and New York: Chapman & Hall/CRC.

Radzki, M. M. (2017a). On Axiom systems of Słupecki for the functionally complete three-valued logic. *Axiomathes, 27*(5), 403–415.

Radzki, M. M. (2017b). On the Rosser-Turquette method of constructing axiom systems for finitely many-valued propositional logics of Łukasiewicz. *Journal of Applied Non-Classical Logics, 27*(4), 27–32.

Rose, A. (1981). Many-valued logics. In E. Agazzi (Ed.), *Modern logic—A survey* (pp. 113–129). Netherlands: Springer.

Rosser, J. B., & Turquette, A. R. (1952). *Many-valued logics*. Amsterdam: North-Holland.

Skala, H. J. (1988). Essays on the history and development of many-valued logics and some related topics. In J. Kacprzyk & M. Fedrizzi (Eds.), *Combining fuzzy imprecision with probabilistic uncertainty in decision making* (pp. 1–9). Berlin and Heidelberg: Springer.

Słupecki, J. (1939). Proof of the axiomatizability of full many-valued systems of propositional calculus. *Comptes rendus des séances de la Société des Sciences et des Lettres de Varsovie, Classe III, 32,* 110–128.

Słupecki, J. (1967). The full three-valued propositional calculus. In S. McCall (Ed.), *Polish logic 1920–1939* (pp. 335–337). Oxford: Clarendon Press.

Słupecki, J., Bryll, G., & Prucnal, T. (1967). Some remarks on three-valued logic of Łukasiewicz. *Studia Logica, XXI,* 45–66.

Tokarz, M. (1974). A method of axiomatization of Łukasiewicz logics. *Bulletin of the Section of Logic, 3,* 21–24.

Tooley, M. (1997). *Time, tense, and causation.* Oxford: Clarendon Press.

Tuziak, R. (1988). An axiomatization of the finite-valued Łukasiewicz calculus. *Studia Logica, 48,* 49–55.

Urquhart, A. (2001). Basic many-valued logic. In D. M. Gabbay & F. Guenthner (Eds.), *Handbook of philosophical logic* (2nd ed., Vol. 2, pp. 249–295). Dordrecht: Kluwer Academic Publishers.

Wajsberg, M. (1967). Axiomatization of the three-valued propositional calculus. In S. McCall (Ed.), *Polish logic 1920–1939* (pp. 264–284). Oxford: Clarendon Press.

Wan, X.-L., & Chen, M.-Y. (2014). The equivalent transformation between non-truth-function and truth-function. In G. Gou & Ch. Liu (Eds.), *Scientific explanation and methodology of science* (pp. 176–211). Singapore: World Scientific Publishing.

Woleński, J. (1989). *Logic and philosophy in the Lvov-Warsaw School: Synthese Library* (Vol. 198). Dordrecht: Kluwer Academic Publishers.

Woleński, J., & Zygmunt, J. (1989). Jerzy Słupecki (1904–1987): Life and work. *Studia Logica, 48,* 401–411.

12

The Methodological Status of Paraphrase in Selected Arguments of Tadeusz Kotarbiński and Kazimierz Ajdukiewicz

Marcin Będkowski

1 Introduction

The operation and concept of paraphrase play a significant role in analytic philosophy, but its role still requires a more profound methodological reflection. In this chapter, we would like to compare the methodological status of paraphrase in selected arguments of Tadeusz Kotarbiński and Kazimierz Ajdukiewicz. We will refer to two examples, namely the argument for the methodological thesis of reism (in Marian Przełęcki's interpretation) and Ajdukiewicz's argument against transcendental idealism.

This chapter is part of a project financed by the Polish National Science Centre; grant number 2015/18/E/HS1/00478.

M. Będkowski (✉)
Faculty of Philosophy and Sociology,
University of Warsaw, Warsaw, Poland
e-mail: mbedkowski@uw.edu.pl

A. Drabarek et al. (eds.), *Interdisciplinary Investigations into the Lvov-Warsaw School*, History of Analytic Philosophy,
https://doi.org/10.1007/978-3-030-24486-6_12

Such formulation of the task is, admittedly, quite ambiguous. According to the well-known distinction introduced by Kazimierz Twardowski (1911), one may consider paraphrase as an action or as a product (of an action). In the former case, a paraphrase is the act or process of restating or rewording, which is a deliberate step in the procedure chosen and followed by a philosopher. In the latter case, it is either a sentence which is a reformulation of the original sentence, or a type of relationship—called 'paraphrastic'—holding between synonymous or nearly synonymous sentences.

In this chapter, we will focus on paraphrase understood as a product of paraphrasing, and we will analyse the role played in philosophical reflection by assertions about the relationship of paraphrase between certain sentences. In other words, we will examine examples of 'arguments from paraphrase' to shed some light on the status of the argumentations of analytic philosophers. We will also discuss some problems posed by the evaluation of such arguments.

To provide further clarification, we will break the initial question about the methodological status of paraphrase into the following questions: What is the place of paraphrastic premises in the structure of selected philosophical arguments? What reasons are offered for those premises? What is the assumed meaning of 'paraphrase', i.e. what conditions must be satisfied so that one sentence can be said to be a paraphrase of another? Can those arguments be regarded as conclusive?

2 Reistic Paraphrase

In a paper entitled 'Argumentacja reisty' ('A Reist's Argumentation', 1984), Przełęcki presents an excellent reconstruction of Kotarbiński's line of reasoning. It is quite an ambitious endeavour, because over the years Kotarbiński's position was repeatedly modified under the influence of subsequent criticisms.

Undoubtedly, one of the most valuable criticisms was put forth by Ajdukiewicz (1930/2006). He distinguishes two theses of reism: a semantic (roughly: 'Every statement about entities other than things is only true if it is reducible to a statement about things') and an ontological one ('Every object is a thing').

As Przełęcki points out, the semantic thesis can be worded as follows (these theses should not be regarded as equivalent):

(T_s) Every meaningful non-reistic sentence is translatable into a reistic sentence.
(T_{sa}) Every true non-reistic sentence is translatable into a reistic sentence.
(T_{sb}) Every meaningful sentence of a non-reistic language is translatable into a sentence of the reistic language.
(T_s^A) Every meaningful non-reistic sentence is translatable into a reistic sentence, on some admissible interpretation. (Przełęcki 1984; cf. Przełęcki 1990)

From the methodological point of view, the main problem which arises when discussing reism concerns logical relations between the theses listed above. The value of the reistic argument depends entirely on which version of the semantic thesis is taken into consideration.

For example, as Przełęcki notices, the ontological thesis entails the semantic thesis T_{sa}, but not T_s. Moreover, although the ontological thesis entails some versions of the semantic thesis, it is the semantic thesis which is the starting point of reistic reasoning. Since our goal is not to consider the whole spectrum of logical dependencies holding between reistic theses, we will not go into detail here. Instead, we will focus on one of the variants of the reistic argument, the one that appears to be the most convincing.

Let us consider the following argument for the methodological thesis of reism (Fig. 1):

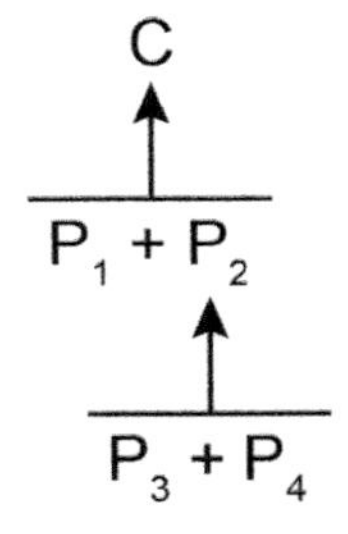

Fig. 1 The argument for the methodological thesis of reism, where 'C' stands for the 'conclusion', and 'P' stands for the 'premise'

(C) Only things may justifiably be said to exist.
(P1) Only those objects may justifiably be said to exist whose existence is assumed by knowledge K.
(P2) Knowledge K assumes only the existence of things.
(P3) Every true non-reistic sentence is translatable into a reistic sentence.
(P4) Reistic sentences assume only the existence of things.

We should introduce several auxiliary definitions. A reistic sentence is a sentence which does not contain any abstract terms, i.e. contains only the names of concrete things. One does assume the existence of objects of a particular type when one asserts a sentence whose subject is a name which refers to an object of that particular type. Traditionally, the Ockham's razor principle states that entities should not be multiplied unnecessarily. 'Knowledge K' is introduced because the semantic thesis needs a relativisation to a specific language. It is natural to consider this language as the language of our knowledge.

Notice that according to Przełęcki:

> The language that Kotarbiński explicitly refers to is, as a matter of course, the Polish language. But the peculiarities of that ethnic language are not essential to the problem in question. What seems essential about the language referred to in thesis (T_s) is the fact that it is a language of our knowledge. This concept plays an important role in the reistic doctrine. The body of our knowledge, identified with a given set of statements K, constitutes the main data of a reistic construction. And it is to this datum that the semantic thesis of reism refers. (Przełęcki 1990: 87)

As we can see, the argument for the methodological thesis of reism is based upon a weak formulation of the semantic thesis and the methodological principle of ontological parsimony. In the opinion of Przełęcki, this argument is quite reasonable and acceptable, being an argument for a thesis weaker than the ontological one. Still, it is of crucial importance whether or not the semantic thesis is true, so we should examine the premise asserting the translatability of particular non-reistic statements in more detail.

As Kotarbiński states, 'the principal justification of reism is naïvely intuitive and based on common induction' (Kotarbiński 1958/1993: 202).

The semantic thesis is justified by the observation that non-reistic sentences, e.g. 'Whiteness is an attribute of snow', may be reformulated and freed from abstract names, e.g. 'Snow is white'. Thus, reism can be seen as a way of eliminating a peculiar manner of speaking in order to restrain its philosophical consequences, such as postulating the existence of entities referred to by non-reistic names.

Of course, common induction, i.e. incomplete inductive reasoning, is, by definition, not valid. Although Kotarbiński shows great ingenuity in rewriting troublesome examples of non-reistic sentences, it is highly questionable whether translations of all non-reistic sentences can be provided. One of such non-translatable sentences is, for example, 'The number of M's is infinite' (Przełęcki 1984: 363). In the next part of this chapter, we will focus on examples of successful and acceptable reistic reformulations, and we will discuss what conditions must be met to consider them reasonable.

Przełęcki's insightful analysis shows that reism struggles with the problem of determining the actual methodological status of such a translation, as well as justification of the semantic thesis of reism. Przełęcki points out that, strictly speaking, the relation between non-reistic and reistic sentences should not be treated as that of translation, but that of paraphrase:

> Here, the concept of translation is applied to expressions whose proper meaning is not literal but metaphorical in the broad sense of the term. Therefore, when saying that a non-reistic sentence Z is translatable into a reistic sentence Z_r, we mean that the proper sense of the proposition Z is not literal but metaphorical and that this sense is identical to the literal sense of Z_r. (Przełęcki 1984: 361, own translation)

Let us notice some difficulties posed by this formulation of the problem. How should we understand the idea that non-reistic sentences are metaphorical? Is 'Whiteness is an attribute of snow' metaphorical in the same way as 'Her eyes are stars' (an example of a typical metaphor) or 'It is raining cats and dogs' (an idiomatic phrase)? In other words, how should one understand 'metaphor in the broad sense'?

Kotarbiński described the origins of reism as follows:

> It began – if memory serves – with doubts regarding properties… […] For example, we hear that qualities belong to things or they reside in things… What is it, this belonging? So we say, but somewhat metaphorically, in a substitutive way. (Kotarbiński 1958/1993: 96, own translation)

Of course, it is easy to regard the sentence 'Whiteness belongs to snow' as metaphorical, because there exists a more basic and literal meaning of the verb 'belong'. However, 'Whiteness is a property of snow' seems to be more problematic. Kotarbiński explicitly argues that the copula 'to be' is used in its literal meaning when it is a sentence-creating functor of two concrete name arguments.

Przełęcki considers this categorial regimentation as arbitrary, but in our opinion, Kotarbiński's position is defensible. He argues that the copula 'to be' has a different meaning depending on the different empirical status of reistic and non-reistic sentences (Kotarbiński 1949/1993).

Since the relation of paraphrase holds between reistic and non-reistic sentences (or the sentences of reistic and non-reistic language), what is the status of this relation? How does it differ from the status of translation? According to Przełęcki, the proper meaning of a non-reistic sentence Z_1 is identical with the literal meaning of a sentence Z_2—i.e. the reistic sentence. Because Z_1 is metaphorical, we should understand the operation of paraphrasing as a kind of literalisation or explication (Przełęcki 1984: 361, own translation).

However, as Jerzy Pelc pointed out, 'only in exceptional cases is the explication of a metaphorical statement (and more generally, of any insufficiently precise statement), limited to giving a single literal and more specific paraphrase' (Pelc 2000: 200, own translation). As an example, Pelc refers to petrified metaphors.

Can non-reistic sentences be treated as such petrified metaphors? Undoubtedly some of them are just that, but sometimes coming up with paraphrases requires a great deal of thought about their interpretation or explication, and some sentences do not have adequate reistic counterparts. The ease of providing paraphrases comes from the fact that in natural language (which interests us in the context of reism) abstract names are derived quite regularly from adjectives and verbs by various grammatical means (e.g. by specialised suffixes).

To shed some light on this issue, in the next section we are going to discuss an argument against reism. In our view, the proper understanding of the relation of paraphrase should offer an answer to this objection.

Anna Wójtowicz, discussing the reduction proposed on the grounds of reism, notes that it is necessary to consider the theory of translation which assigns reistic sentences to non-reistic ones:

> On the basis of what theory of translation (T) does one translate natural language into a reistic sublanguage? If the translation is understood as a translation *salva veritate*, theory T must assign true sentences containing apparent names to relevant true sentences which do not contain such names. However, the assertion that a certain sentence containing apparent name A is true—e.g. "A has the property P"—is related to the assertion that name A designates something and that, in fact, the referent of this name has the property P. On the basis of theory T, name A is not an apparent name. In other words, a natural language containing apparent names is translated into a reistic sublanguage only on the basis of a theory in which apparent names are no longer apparent. (Wójtowicz 2007: 26, own translation)

Is this criticism valid in the light of what was said in the previous section about the status of paraphrase? According to Kotarbiński, sentences of the type: 'A is B' are true only if there exists an object that A refers to. However, we should remember that this condition is limited to literal expressions, i.e. those which contain the copula 'to be' in its primary meaning, so it does not include substitutive abbreviations:

> [...] an expression of the type: "P is a denominatum of a name N" is a substitutive abbreviation. Were it not that – were it to be taken literally, its acceptance would result in the acceptance of the existence of an object P concerning which it could be claimed that it is a denominatum of N. For this expression is a special case of the schema 'A is B', and for a statement constructed according to that schema to be true, the existence of an object A would be required. But those expressions which are substitutive

> abbreviations do not comply with that rule – if we bear in mind what they are substitutive abbreviations of. If we say, e.g., that "a shaving knife is a denominatum of the name 'razor'" we merely claim that "whoever understands the name 'razor' has a shaving knife in mind", and if we say that a person has a shaving knife in mind we do not claim indirectly that any shaving knife exists; likewise, when we say that a person has a millegenarian in mind we do not claim that any such old human being exists. (Kotarbiński 1963/1971: 63-64)

As we can see, truth-values of non-reistic sentences depend solely on truth-values of their reistic counterparts. Thus, an assignment between reistic and non-reistic sentences is a relation of paraphrase, and we should abandon the condition that the theory of translation should identify sentences according to their interchangeability *salva veritate* or *salva sensu*. According to Kotarbiński, non-reistic sentences have no autonomous (independent from their paraphrases) truth conditions.

Let us consider the rationale for this view. Non-reistic sentences are considered metaphorical not only due to their vagueness but also their origins. From the reistic point of view, a non-reistic sentence (e.g. 'Love is suffering') is an abbreviation and a substitute for a reistic sentence (e.g. 'Everyone who loves suffers.') The former sentence—containing apparent names—can be treated as meaningful only if there exists an acceptable paraphrase which does not include apparent names. Recall that Przełęcki calls the operation which enables us to construct such reformulations 'paraphrase'; it is more akin, however, to literalisation or explication, but also to decoding or deciphering.

Let us consider a simple example of deciphering. The following sequence of letters 'WKHFDW LV RQWKHPDW' is an equivalent of 'The cat is on the mat', according to rules of Caesar cipher and the transformation based on a right shift of three which assigns: 'A' to 'D', 'B' to 'E', 'C' to 'F', and so on. We can now ask whether the encrypted sentence is meaningful. It seems so, although this sequence of letters is not even a proper expression of English language (or any natural language).

However, it has a meaningful counterpart (unlike, for example, the sequence 'WKKFDD LV RQWKKPDD'). If it is meaningful, it is so due to the transformation which reveals its proper linguistic form.

We think that the reist wants to claim something similar regarding non-reistic sentences: they are meaningful, but only due to some transformations revealing their real form and meaning.

Like deciphering, the operation of reistic paraphrase is (or at least should be) purely syntactical and based on the correlation between the syntactical schemes of reistic and non-reistic expressions, e.g. 'X-ness belongs to Y'—'Y is X' (cf. Smaby 1971). The reist's goal is to provide such correlations for all expressions containing abstract names, i.e. indicating all possible syntactical schemes with apparent names. If the sentences being the results of paraphrasing (simply speaking, paraphrases) are meaningful, then we can also accept their non-reistic analogs as meaningful.

All thing considered, we should demand from a theory of reductive paraphrase (translation) to be coherent with the target language in respect of qualities which are affected by this reduction. In order to eliminate a particular quality from sentences of a language, we cannot use a theory of translation which exploits this quality. In particular, the theory of reistic paraphrase has to be reistic itself; it cannot take apparent names to be genuine names.

It is worth noting that the operation of reductive paraphrase is connected with explication. It seems that some, if not most, schemes of non-reistic sentences can have more than one reistic equivalent (e.g. 'Love is suffering' can be understood as 'Everyone who loves suffers' or 'Everyone who is loved suffers', etc.) In some cases, explication can be a demanding endeavour. This makes the claim that non-reistic sentences are metaphorical (in a broad sense) fairly acceptable. At the same time, however, it complicates the issue of justification of reistic paraphrases.

3 Explicative Paraphrase

In this section, we will consider a different type of argument from paraphrase. So far, we have discussed the reistic paraphrase, which exemplifies the type that can be called more generally 'reductive'. Now we will direct our attention to one which can be called 'explicative'.

Ajdukiewicz is considered to be the author of a so-called paraphrase method. In our opinion, speaking of a paraphrase method is misleading (there is in fact a number of approaches at work here). Still, there is no doubt that in his works, Ajdukiewicz explicitly uses the notion of paraphrase (cf. Ajdukiewicz 1934b/1978, 1934c/1978, 1937/1978). At this point, we will limit ourselves to considering arguments presented in his article 'Problemat transcendentalnego idealismu w sformułowaniu semantycznym' ('A Semantic Formulation of the Problem of Transcendental Idealism'; Ajdukiewicz 1937/2006).

In his article, Ajdukiewicz tries to refute the main thesis of transcendental idealism, namely that 'Reality is nothing but a correlate of the transcendental subject'. However, this thesis—as well as the crucial terms of idealistic philosophy—'lack clear-cut meaning'. Therefore, Ajdukiewicz's main effort is to restate this thesis in a more precise language—which in this case is the language of logical semantics—using the concepts of thesis, theorem, deductive system, completeness and so on.

Next, in order to disprove the thesis of idealism, Ajdukiewicz refers to Kurt Gödel's first incompleteness theorem. In short, Ajdukiewicz's argument consists essentially in reducing the thesis of idealism to a common denominator with the results of metalogic. This is what Jan Woleński metaphorically describes as 'immersing one theory in another' (Woleński 1987: 99).

The argument for the thesis of idealism in the semantic formulation proposed by Ajdukiewicz can be reconstructed as follows (Fig. 2):

(C) The system of the language of natural sciences is complete.

(P1) A deductive system is complete if and only if for every statement of the language of the system, either the statement or its negation is a theorem of this system.

(P2) In the system of the language of natural sciences, only those statements are true which are theorems of this language.

(P3) True are all and only those judgments in the logical sense which are dictated by transcendental norms.

(P4) Judgements in the logical sense correspond to theorems of the deductive system.

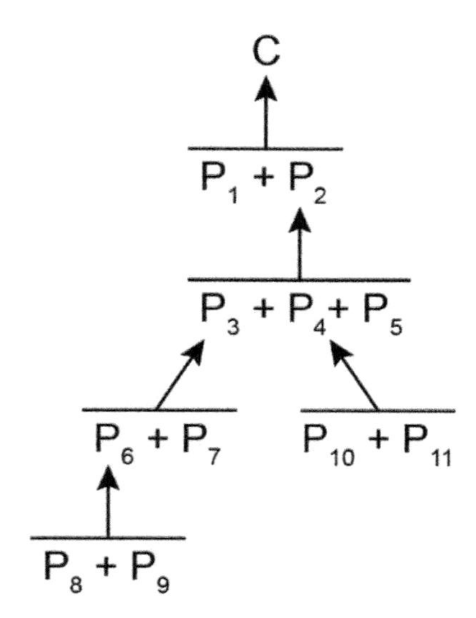

Fig. 2 The argument for the thesis of idealism in the semantic formulation proposed by Ajdukiewicz, where 'C' stands for the 'conclusion', and 'P' stands for the 'premise'

(P5) The system under consideration is the system of (the language of) natural sciences.
(P6) Reality is a correlate of the set of judgments (in the logical sense) dictated by transcendental norms.
(P7) Every judgment in the logical sense is true if and only if it correlates with a fragment of reality.
(P8) Reality is nothing but a correlate of the transcendental subject.
(P9) The transcendental subject is the set of judgments (in the logical sense) dictated by transcendental norms.
(P10) A relation of paraphrase holds between transcendental norms and rules of direct consequence.
(P11) Theorems of the language (of the deductive system) are sentences dictated by the rules of direct consequence.

The purpose of paraphrase is to establish correspondence between the following pairs of terms and theses:

- 'transcendental norms'—'linguistic rules of direct consequence',
- 'transcendental subject'—'set of true statements',
- 'judgment dictated by transcendental norms'—'theorem of the language (of the deductive system)',
- 'Reality is nothing but a correlate of the transcendental subject.'—'The system of the language of natural sciences is complete.'

It is worth noting that in Ajdukiewicz's article the arguments of the relation of paraphrase are: the position of idealism, the concept of the norm, and the thesis of idealism. It is a little peculiar because the paraphrastic relation is usually understood as a relation holding between sentences. So, it is natural to apply the concept of paraphrase to theses or theorems, while it is quite odd to speak about paraphrases of concepts. Although Ajdukiewicz could defend himself saying that he talks about paraphrases of concepts for the sake of brevity or metonymically, it seems that in the case of concepts, one should talk about their explication instead.

Once again, we should consider the relation of paraphrase in comparison to the relation of translation. Ajdukiewicz presented two different conceptions of translation. The first account is formulated on the basis of his renowned directival theory of meaning (cf. Ajdukiewicz 1931/2006, 1934/2006; Grabarczyk 2017). In this theory, Ajdukiewicz explicates the concept of meaning in terms of meaning-rules. They require each user of the language to accept certain sentences of this language (in certain situations). If one accepts certain sets of sentences containing a given expression, then he or she can be said to know its meaning (Ajdukiewicz 1934a/1978: 44). Ajdukiewicz distinguishes three types of rules: axiomatic, deductive and empirical. Let us briefly characterise those types.

Regarding deductive (inferential) rules, Ajdukiewicz states:

> Only that person connects with expressions of the English language the meaning coordinated with them in that language who is prepared to accept the sentence 'B' as soon as he accepts sentences of the form: 'if A, then B' and 'A'. (Ajdukiewicz 1934a/1978: 43)

In other words, the deductive rule of meaning correlates two sentences of different types: a premise and a conclusion. It is expected of a user of language to draw conclusions from premises.

Axiomatic rules are described as follows:

> Turning to what we call 'axiomatic' rules of meaning, we find these best illustrated in the languages of axiom-systems. There, again, an unconditional demand is made of anybody who wants to use the words of an axiom-system in the meaning coordinated with them by the language of the system: that the person be prepared to accept without further ado the sentences set up as the axioms of the system. (Ajdukiewicz 1934a/1978: 45–46)

Of course, axiomatic rules are not limited to axiom-systems. We can find them in everyday language, e.g. everyone who knows the meaning of the words 'every' and 'is' in English accepts the sentence 'Every dog is a dog'.

With respect to empirical rules, Ajdukiewicz states:

> These rules are characterised by the fact that the situations they involve (in which readiness to accept a sentence is required by the meaning-specification of the language) consist either exclusively or partly in experiencing a perception. One apparently proceeds according to such a rule when they accept the sentence 'It hurts' upon experiencing a toothache. (Ajdukiewicz 1934a/1978: 46)

Regarding the concept of translation, Ajdukiewicz considers two cases. The first case is the identity of meaning—or synonymity—of two expressions of the same language S. The necessary condition for synonymity is as follows:

> If in S expressions A and A' have the same meaning, then in the total scope of the meaning-rules of S expressions A and A' behave as isotopes, i.e. the total scope of the meaning-rules is not altered if, in all its members, occurrences of A are replaced by A', and those of A' by A. (Ajdukiewicz 1934a/1978: 53)

Next, Ajdukiewicz elaborates on substituting expressions in three kinds of meaning-rules. In short, it means that we can substitute synonymous expressions across the rules and we should accept sentences or inferences whose acceptance is required by rules obtained by replacing A with A'.

Now, let us turn to the second case, in which translation is understood as the identity of meaning of two expressions of different languages:

> If expression A in language S has the same meaning as expression A' in language S', then we call expression A' a 'translation' of A from language S into language S'. The relation 'is a translation of' is reflexive, symmetric and transitive. (Ajdukiewicz 1934a/1978: 54)

In consequence:

> If A' is a translation of A from language S into language S'; and if in S, moreover, A has direct meaning-relations with expressions A_1, A_2, ..., A_n, which latter expressions likewise have translations $A_{1'}$, $A_{2'}$, ..., $A_{n'}$ into language S'; then in language S', expression A' must have direct meaning-relations with $A_{1'}$, $A_{2'}$, ..., $A_{n'}$ which severally are analogous to those A has with A_1, A_2, ..., A_n in language S. (Ajdukiewicz 1934a/1978: 55)

According to this view, since the meaning-rules of a language concern not only expressions (Ajdukiewicz 1934a/1978: 48), but syntactical forms as well, expressions and syntactic forms are both translatable:

> The syntactic forms admitted by a language – forms of expressions which comprise more than one word – are determined by the rules of syntax of the language and are as peculiar to that language as its vocabulary. Thus, these forms undergo translation just as words do. (Ajdukiewicz 1934a/1978: 55)

So, generally speaking, if an expression of a language is to be translatable into another language, the meaning-rules and syntactic form of the former must be coordinated with analogous meaning-rules and syntactic form of the latter.

In his second theory, much later, Ajdukiewicz does not refer to the concept of meaning; on the contrary, he explicitly avoids it:

> To define the notion of translation we shall not refer to the notion of meaning, in particular, we shall not say that an expression A is a translation of an expression B if both these expressions have the same meaning; such a definition would burden the notion of translation with all those obscurities which are connected with the term 'meaning'. Our definition of translation will be based on the notions of denotation and of syntactic position. (Ajdukiewicz 1967/1978: 328)

It is worth noting that from the logical point of view, expressions A and B can belong to the same language, i.e. it is not only about two expressions belonging to different languages that we may say that one is a translation of another (Ajdukiewicz 1967/1978: 328). Ajdukiewicz also draws attention to the fact that we can consider translations to vary in degrees of precision, i.e. to be more or less literal and adequate. Having said that, he presents the following definition of literal translation:

> We say that an expression A is a literal translation of an expression B if, and only if, after the resolution of all the abbreviations which they contain, they are transformed into two abbreviation-free expressions A' and B', such that there is a one-to-one correspondence between all the constituents of one expression and the constituents of the other expression; moreover, the constituents between which such correspondence is established, (1) occupy in A' and B', respectively, the same syntactic positions, and (2) are reciprocally equivalent, i.e., denote the same object. (Ajdukiewicz 1967/1978: 328)

And what about less literal translation?

> [...] an expression A is an nth degree translation of an expression B if, and only if, after the resolution of all the abbreviations which they contain they are transformed into two abbreviation-free expressions A' and B', such that there is a one-to-one correspondence between all the constituents of the nth and lower orders of each expression; moreover, if the constituents between which such correspondence is established (1), occupy in A' and B', respectively, the same syntactic positions and (2) are reciprocally equivalent, i.e. denote the same object. (Ajdukiewicz 1967/1978: 328)

Of course, literal translation is a particular case of nth degree translation, namely the translation which reaches the highest degree of precision.

It should be evident to us that Ajdukiewicz does not treat paraphrastic relation as a relation of the identity of meaning holding between sentences. In other words, his paraphrase of the thesis of idealism cannot be interpreted even as a zero-degree translation (i.e. lowest degree of precision). In fact, the operation conducted by Ajdukiewicz is a bold semantic experiment, as it engages both the vague language of traditional epistemology and the language of semantic epistemology.

As we can easily see, Ajdukiewicz abandoned the idea of translating syntactic forms for the idea of correspondence between syntactic positions. Although this idea was elaborated on in one of his last works, it had been partially developed in (Ajdukiewicz 1934b/2006). Nevertheless, a one-to-one correspondence between constituents of the thesis of idealism is easy to demonstrate. Following Jacek J. Jadacki, we can present this correspondence as follows in Table 1.

Taking this into account, it is not surprising that Jadacki puts the condition of sameness (isomorphism) of structures into what he offers as the definition of paraphrase in Ajdukiewicz's tradition:

Paraphrase (of *phrases*—especially sentences—or *texts*), i.e. structural travesty

Table 1 A one-to-one correspondence between constituents of the thesis of idealism and its paraphrase

The real	world	is	exclusively	a correlate of the transcendental subject, i.e. judgments dictated by the transcendental norm
True	sentences of the scientific language	are	exclusively	the theses of this language, i.e. judgments dictated by the deductive rules of this language

$$\forall\alpha\forall\beta\ (\alpha \text{ is a paraphrase of } \beta) \Leftrightarrow \{[(E\alpha \neq E\beta)_{\text{isomorphic structure}} \wedge (S\alpha \neq S\beta)] \wedge [(K\alpha \neq K\beta)_{\text{as to precision}} \wedge (D\alpha = D\beta)]\}.$$ (Jadacki 1995: 151)

where: '*E*' symbolises 'elements', '*S*'—'structure', '*K*'—'connotation', '*D*'—'denotation of the sign'.

It is worth noting that the paraphrase of the thesis of idealism is based on an approach similar to Rudolf Carnap's theory of explication and intensional isomorphism (Carnap 1947). Corresponding elements of the thesis of idealism and its semantic formulation (transcendental norm—linguistic rules of direct consequence, transcendental subject—set of true statements, and so on) are arguments of the kind of relation which holds between *explicandum* and *explicans*.

Jadacki describes this relation as governed by the following conditions: *explicandum* and *explicans* have the same denotation, but they have different connotations; more specifically, the connotation of *explicans* is more precise than the connotation of *explicandum* (Jadacki 1995).

As argued by Katarzyna Paprzycka, it is not always the case that *explicandum* and *explicans* have the same denotation but different connotations. It is often the case that they differ in both aspects. One of the prototypical examples of explication is Alfred Tarski's explication of the notion of truth. It can hardly be claimed that the *explicandum* and the *explicans* have the same denotations in this case. Instead, the *explicans* satisfies different conditions, formulated by Tarski, based on intuitions related to the nebulous traditional notion of truth.[1]

Now we have to face the problem of justification for paraphrase. In the case of explicative paraphrases, it is clear that they are founded on an operation similar to explication as understood by Carnap (although we should remember that it predates Carnap's concept, which was developed in the 1950s). Of course, transformation of the original problem has the advantage of precision, being well grounded

[1]We should emphasise that the main theoretical function of explication is to specify vague concepts (cf. Brożek 2015: 37).

in scientific context. But the question arises: what are the grounds for Ajdukiewicz's paraphrases? To what extent do they correspond to the original formulation and intentions of Heinrich Rickert?

As Jerzy Giedymin states:

> [Ajdukiewicz] provided semantical paraphrases of the doctrines of subjective and logical idealism and a critique of those doctrines based on his 'translation', though he was careful to admit that he could not be certain of the adequacy of the translation and that the authors of those doctrines would in all probability not regard his translations as adequate. (Giedymin 1978: XXI)

Ajdukiewicz devoted a lot of space to a discussion of the adequacy of his paraphrases, while emphasising that the language of idealism needed to be clarified. He discussed the way idealists understood the nature of a transcendental subject, showing how the concept changed over time. Initially, idealists focused on its ontological nature and relation to individual consciousness; later on, they conceived it in a more abstract and functional way.

Nevertheless, the justification of the paraphrastic premise is pragmatic and hermeneutic—it is based on an interpretation of the philosophy of idealism. Even if it is not a good interpretation—unacceptable for idealists—it shifts to them the burden of proof.

4 Conclusion

Let us recapitulate the most important points. Both Kotarbiński and Ajdukiewicz referred in their philosophical inquiries to so-called arguments from paraphrase. These arguments assert as one of their premises that there is a paraphrastic relation holding between certain expressions. This type of argumentation is quite common in analytic philosophy, perhaps even essential for it. The methodological status of this operation and relation requires more reflection, however.

In order to shed some light on the status of paraphrase, we discussed two examples. In the arguments we analysed, paraphrases served to justify a methodological (or ontological) thesis and an epistemological thesis, respectively. Kotarbiński's argumentation is of a reductive type; it consists in recalling reasons for a certain methodological programme or for a modest ontology. Ajdukiewicz's argument, on the other hand, is aimed primarily at clarifying certain traditional philosophical categories and challenge theses expressed in these terms.

Various objections challenge the value of their arguments (cf. Wójtowicz 2003; Jadacki 1995: 151–152). In a way, these objections indicate the vagueness and ambiguity of the notion of paraphrase. In the case of reistic argumentation, it is difficult to provide a theory which is consistent with reism. A reist should be able to formulate a reistic theory of translation, and in particular should not assume non-reistic expressions to be primarily meaningful. It appears that this theory should match non-reistic and reistic sentences, the sole basis of their syntactic structure.

The metaphorical nature of non-reistic sentences seems to be another interesting issue. Presumably, reistic paraphrases have not only a reductive, but also an explanatory function—they elucidate the meaning of non-reistic sentences. Due to their syntactic nature, reistic paraphrases are close in character to deciphering. This issue, however, requires further elaboration.

In the case of Ajdukiewicz, paraphrase has an explicative nature. He tries to reformulate the position of idealism, expressed in the thesis of idealism. To this end, he establishes a correlation between expressions belonging to the idealist language and expressions belonging to the language of semantic epistemology. By using metalogical theorems, it is possible to demonstrate the falsehood of the thesis of idealism in its semantic formulation. The aim of this procedure is clear: it is about extrapolating scientific results onto the grounds of philosophy. In other words: it is about drawing consequences of an interpretative nature.

Another difficulty is the problem of justification of paraphrases. In the case of a reist, justification for a semantic thesis is naïvely intuitive and based on common induction. On the basis of relatively simple examples, a reist sets up general principles of syntactic translation.

Ajdukiewicz refers to a certain hermeneutic procedure, whose principles are not strictly defined. It aims to elicit the most important functions of the analysed concepts. Intuitions connected with those vague concepts can be expressed in the form of conditions that should be met by a correct analysis, while other, potentially conflicting, intuitions can be ignored. At least, until one can express them with precision.

References

Ajdukiewicz, K. (1930/2006). Reizm. In K. Ajdukiewicz, *Język i poznanie, vol. 1, Wybór pism z lat 1920–1939* (pp. 79–101). Warszawa: Wydawnictwo Naukowe PWN.

Ajdukiewicz, K. (1931/2006). O znaczeniu wyrażeń. In K. Ajdukiewicz, *Język i poznanie, vol. 1, Wybór pism z lat 1920–1939* (pp. 102–136). Warszawa: Wydawnictwo Naukowe PWN. English translation: Ajdukiewicz, K. (1931/1978). On the meaning of expressions. In J. Giedymin (Ed.), *The scientific world-perspective and other essays, 1931–1963* (pp. 1–34). Dordrecht and Boston: D. Reidel Publishing Company.

Ajdukiewicz, K. (1934a/2006). Język i znaczenie. In K. Ajdukiewicz, *Język i poznanie, vol. 1, Wybór pism z lat 1920–1939* (pp. 145–174). Warszawa: Wydawnictwo Naukowe PWN. English translation: Ajdukiewicz, K. (1934a/1978). Language and meaning. In J. Giedymin (Ed.), *The scientific world-perspective and other essays, 1931–1963* (pp. 35–66). Dordrecht and Boston: D. Reidel Publishing Company.

Ajdukiewicz, K. (1934b/2006). O stosowalności czystej logiki do zagadnień filozoficznych. In K. Ajdukiewicz, *Język i poznanie, vol. 1, Wybór pism z lat 1920–1939* (pp. 211–214). Warszawa: Wydawnictwo Naukowe PWN. English translation: Ajdukiewicz, K. (1934b/1978). On the applicability of pure logic to philosophical problems. In J. Giedymin (Ed.), *The scientific world-perspective and other essays, 1931–1963* (pp. 90–94). Dordrecht and Boston: D. Reidel Publishing Company.

Ajdukiewicz, K. (1934c/2006). W sprawie «uniwersaliów». In K. Ajdukiewicz, *Język i poznanie, vol. 1, Wybór pism z lat 1920–1939* (pp. 196–210). Warszawa: Wydawnictwo Naukowe PWN. English translation: Ajdukiewicz, K. (1934c/1978). On the problem of universals. In J. Giedymin (Ed.), *The scientific world-perspective and other essays, 1931–1963* (pp. 95–110). Dordrecht and Boston: D. Reidel Publishing Company.

Ajdukiewicz, K. (1937/2006). Problemat transcendentalnego idealizmu w sformułowaniu semantycznym. In K. Ajdukiewicz, *Język i poznanie, vol. 1, Wybór pism z lat 1920–1939* (pp. 264–277). Warszawa: Wydawnictwo Naukowe PWN. English translation: Ajdukiewicz, K. (1937/1978). A semantical version of the problem of transcendental idealism. In J. Giedymin (Ed.), *The scientific world-perspective and other essays, 1931–1963* (pp. 264–277). Dordrecht and Boston: D. Reidel Publishing Company.

Ajdukiewicz, K. (1967/1978). Intensional Expressions. In J. Giedymin (Ed.), *The scientific world-perspective and other essays, 1931–1963* (pp. 320–347). Dordrecht and Boston: D. Reidel Publishing Company.

Brożek, A. (2015). O naturalizacji dyscyplin naukowych. *Filo-Sofija, 29,* 29–42.

Carnap, R. (1947). *Meaning and necessity: A study in semantics and modal logic.* Chicago: Chicago University Press.

Giedymin, J. (1978). Editor's introduction: In radical conventionalism, its background and evolution: Poincaré, LeRoy, Ajdukiewicz. In J. Giedymin (Ed.), *The scientific world-perspective and other essays, 1931–1963* (pp. XIX–LIII). Dordrecht and Boston: D. Reidel Publishing Company.

Grabarczyk, P. (2017). Directival theory of meaning resurrected. *Studia Semiotyczne, XXXI/1,* 24–44.

Jadacki, J. J. (1992). Definicja, eksplikacja i parafraza w tradycji Ajdukiewiczowskiej. In H. Zelnik (Ed.), Fragmenty filozoficzne ofiarowane§ Henrykowi Hiżowi w siedemdziesiątą piątą rocznicę urodzin. Biblioteka Myśli Semiotycznej 23 (pp. 28–40). Warszawa: Zakład Semiotyki Logicznej Uniwersytetu Warszawskiego, Znak Język Rzeczywistość, Polskie Towarzystwo Semiotyczne. English translation: Jadacki, J. J. (1995). Definition, explication and paraphrase in the Ajdukiewiczian Tradition. In V. Sinisi & J. Woleński (Eds.), The heritage of Kazimierz Ajdukiewicz. Poznań studies in the philosophy of the sciences and the humanities 40 (pp. 139–152). Amsterdam and Atlanta: Rodopi.

Kotarbiński, T. (1949/1993). O postawie reistycznej, czyli konkretystycznej. In W. Gasparski, D. Miller, & M. Adamczewska (Eds.), Dzieła wszystkie,vol. 1, Ontologia, teoria poznania i metodologia nauk, (pp. 149–158). Wrocław, Warszawa, and Kraków: Zakład Narodowy im. Ossolińskich. English translation: Kotarbiński, T., The reisticor concretistic approach. In J. Pelc (Ed.), Semiotics in Poland. 1894–1969. Synthese Library 119 (pp. 40–51). Dordrecht: D. Reidel.

Kotarbiński, T. (1958/1993). Fazy rozwojowe konkretyzmu. In W. Gasparski, D. Miller, & M. Adamczewska (Eds.), Dzieła wszystkie, vol. 1, Ontologia, teoria poznania i metodologia nauk (pp. 196–205). Wrocław, Warszawa, and Kraków: Zakład Narodowy im. Ossolińskich.

Kotarbiński, T. (1963/1993). Spór o desygnat. In W. Gasparski, D. Miller, & M. Adamczewska (Eds.), Dzieła wszystkie, vol. 1, Ontologia, teoria poznania i metodologia nauk (pp. 491–495). Wrocław, Warszawa, and Kraków: Zakład Narodowy im. Ossolińskich. English translation: Kotarbiński T. (1963/1971). The Controversy over designata. In J. Pelc (Ed.), Semiotics in Poland. 1894–1969. Synthese Library 119 (pp. 59–64). Dordrecht: D. Reidel.

Pelc, J. (2000). Myśli o języku humanistyki. In J. Pelc (Ed.), *Język współczesnej humanistyki* (pp. 171–206). Warszawa: Znak Język Rzeczywistość, Polskie Towarzystwo Semiotyczne.

Przełęcki, M. (1984). Argumentacja reisty. *Studia Filozoficzne, 5*(22), 5–22.

Przełęcki, M. (1990). Semantic Reasons for Ontological Statements: The Argumentation of a Reist. In J. Woleński (Ed.), *Kotarbinski: Logic, semantics and ontology* (pp. 85–96). Dordrecht: Kluwer.

Smaby, R. M. (1971). Paraphrase Grammars. Formal Linguistics Series 2. Dordrecht: Reidel.

Twardowski, K. (1911). O czynnościach i wytworach. Kilka uwag z pogranicza psychologii, gramatyki i logiki. Lwów: Księgarnia Gubrynowicza i Syna.

Woleński, J. (1987). Metamatematyka i filozofia. In M. Heller, A. Michalik, & J. Życiński (Eds.), *Filozofować w kontekście nauki* (pp. 96–104). Kraków: Polskie Towarzystwo Teologiczne.

Wójtowicz, A. (2003). Kilka uwag o argumencie 'z przekładalności'. In W. Krajewski & W. Strawiński (Eds.), *Odkrycie naukowe i inne zagadnienia współczesnej filozofii nauki* (pp. 169–177). Warszawa: Wydawnictwo Naukowe Semper.

Wójtowicz, A. (2007). *Znaczenie nazw a znaczenie zdań. W obronie ontologii sytuacji*. Warszawa: Wydawnictwo Naukowe Semper.

13

The Metaphilosophical Views of Zygmunt Zawirski Against the Background of Contemporary Discussions on Interdisciplinarity in Science

Jarosław Maciej Janowski

1 Introduction

This chapter will focus on two main issues. The first one can be outlined by reference to a quote from Jadwiga Mizińska:

> Calls for interdisciplinarity are a reaction to the adverse effects of excessive specialization. [...] It might appear that the proper reaction to the excessive shredding of reality (reflected in the division of the "kingdom of science" into ever smaller "principalities") should be their reunification. The question, however, is on what terms should it be performed? Who and on what grounds should decide? Under whose rule should this unification take place? (Mizińska 2012: 71–72)

J. M. Janowski (✉)
Institute of Philosophy and Sociology,
The Maria Grzegorzewska University, Warsaw, Poland
e-mail: jjanowski@aps.edu.pl

A. Drabarek et al. (eds.), *Interdisciplinary Investigations into the Lvov-Warsaw School*, History of Analytic Philosophy,
https://doi.org/10.1007/978-3-030-24486-6_13

In what follows, we will refer to the discussion, resurfacing every now and again in ever new versions, concerning the postulate of interdisciplinary research in science. The term 'interdisciplinarity' itself has appeared relatively recently—it had not been around until the second half of the twentieth century (Hejmej 2007: 38). Quite soon, however, the range of problems related to it grew to enormous dimensions, and the term interdisciplinarity itself turned out to be an ambiguous notion. A number of concepts related to that of interdisciplinarity soon appeared as well: multidisciplinarity, transdisciplinarity, unidisciplinarity, pluridisciplinarity, or syndisciplinarity (Poczobut 2012; Walczak 2016).

One of the Polish researchers, who studies this phenomenon in the humanities, cites the opinion of the American scholar Julie Thompson Klein, who says that 'interdisciplinarity has been variously defined in this century: as a methodology, a concept, a process, a way of thinking, a philosophy' (Klein 1990: 196; Hejmej 2007: 38). In the Polish academic circles, the last 'great' debate on the issue of interdisciplinarity began in 2009 during a panel discussion on 'Interdisciplinarity in Science as Truism, Alibi and Challenge' during a convention of the Polish Academy of Sciences (PAN) Committee of Pedagogical Sciences, finalised with a monograph entitled 'Interdisciplinarily on Interdisciplinarity. Between Idea and Practice' (Chmielewski et al. 2012).[1] At the same time, it appears that studies on interdisciplinarity in philosophy are treated as marginal to reflections concerning other fields of knowledge.[2] Which is why we will try to refer the above

---

[1]This publication is probably one of the main sources of information in Polish on issues related to the problem of interdisciplinarity. Foreign publications include i.a.: Klein (1990), Fordeman (2010).

[2]We wish to stress, however, that it is not our intention to diminish the monograph in any way. We agree with Leszek Korporowicz who says in the editor's review that the monograph 'depicts in the plurality of issues discussed [...] their complexity, diversity, and the sometimes not apparent implications of the problems, providing the reader with a very broad horizon for their possible understanding, their interrelations and developmental dynamics in the context of new challenges, needs, and possibilities', and which 'will be an absolutely unique publication in Polish scientific literature, representing a holistic approach and playing an analytical, but inspiring role, making the issues discussed far from anachronism and particularity of viewpoint' (Chmielewski et al. 2012: cover).

problems related to interdisciplinarity to metaphilosophical postulates put forward by the Lvov-Warsaw School, particularly to proposals related to the concept of so-called scientific philosophy.

A general outline of the metaphilosophical attitude characteristic of representatives of the Lvov-Warsaw School can be found in the following quote from Jan Woleński:

> The philosopher type of the Lvov-Warsaw School paradigm is comparatively easy to reconstruct. A philosopher should be educated in philosophy and trained in some empirical discipline, preferably in the field of mathematics or natural sciences. Philosophical education should be both historical and systematic, with particular emphasis on being up to date with developments in contemporary philosophy. (Woleński 1985: 32)

We will then look at the views of Zygmunt Zawirski, whose studies were also focused on issues related to the philosophy of nature. To this end, we will refer to his postulates concerning the practice of scientific philosophy, and the relationship between science and metaphysics. Finally, we will refer to reflections on the issue of time,[3] and in this context we will try to determine whether Zawirski's philosophical reflections and his metaphilosophical views correspond to today's postulates concerning interdisciplinary studies in science, including philosophy.

According to Robert Poczobut (2012: 39–40), the problem of interdisciplinarity appears in reflections on methodology and the philosophy of science in relation to analyses concerning the process of evolution in sciences. It includes 'two complementary tendencies, referred to as divergence and convergence'. The first one consists in an increasing specialization of studies, which leads to progressive fragmentation of particular fields of knowledge. The other tendency, referred to as

[3] It should be noted here that the problem of time is one of the most essential questions considered by philosophy, which has yet to be provided with a satisfactory solution. Zawirski's analyses related to the evolution of the concept of time discussed further on in this chapter are considered by Woleński to be one of the most thorough ones in the then available literature of the subject. And even though he admits that Zawirski 'does not offer any theory of time, his [publication on the subject] contains a great number of exquisitely subtle comments on most of the disputed problems of the philosophy of time' (Woleński 1985: 229).

convergence, aims at integrating and unifying various areas and fields of study. Being the outcome of diverse relationships between particular disciplines, it allows for the 'creation of multi-disciplines, inter-disciplines, and trans-disciplinary programmes, as well as attempts at partial reduction of certain theories and disciplines of science'. At the same time, according to Andrzej Hejmej (2007: 36–38): 'various attempts are made at implementing and explaining [interdisciplinarity] from the perspective of different research disciplines'. Moreover, 'every dispute about interdisciplinarity is a dispute about the rules of existence and condition of particular disciplines', and is thus a dispute 'which necessarily places the issue of status and boundaries of particular research disciplines in the centre of attention'. This last problem has been extensively discussed by representatives of the Lvov-Warsaw School of philosophy.

2 The Metaphilosophical Views of Zawirski

In the introduction to a monograph on philosophical interpretations of scientific facts, Józef Turek (2009: 5–12) mentioned two types of standpoints on the relationship between philosophy and empirical sciences. The first type refers to autonomous philosophies, i.e. such as are methodologically and epistemologically independent from empirical sciences. These in turn divide into philosophies open to the achievements of empirical sciences, and philosophies which disregard and are generally closed to them. The other standpoint refers to non-autonomous philosophies which are supposed to somehow correspond to empirical sciences. Particularly with regard to the latter, but also to philosophies which are autonomous but open to the achievements of empirical sciences, especially natural sciences, problems arise with moving from scientific to philosophical knowledge. 'The problem has a long history and in its theoretical dimension has most often been, and still is, reduced to a discussion about practical ways of scientifying philosophy, or turning it into a so-called scientific philosophy' (Turek 2009: 5).

'Scientific philosophy' was first discussed in the nineteenth century in view of the swift progress taking place in natural sciences. 'Positivists claimed that empirical sciences had spun off from philosophy, eventually leaving it without its proper subject of study. One of the first protagonists of a scientific concept of philosophy in the nineteenth century was Auguste Comte; subsequently, it was embraced by logical positivists of the Vienna Circle' (Duchliński 2014: 188). Discussions around this subject were also held in the Lvov-Warsaw School. In his address delivered on the occasion of the opening of the Polish Philosophical Society in Lvov, Kazimierz Twardowski said:

> The only dogma of the [Polish Philosophical] Society will be the conviction that dogmatism is the greatest enemy of any scientific work. [...] We want all directions pursued and all views held in our Society to be aimed at one goal: discovering the truth. The road which leads there [is] scientific criticism. (Twardowski 1904: 241–242)

Commenting upon these words of the founder of the Lvov-Warsaw School, Woleński says:

> Twardowski—and his followers—did not impose any particular philosophical doctrine upon his students; he intended to teach them a clear and critical way of thinking in the first place. It appears that the philosophical minimalism Twardowski announced *ex officio* was a well-thought element of a global concept of creating philosophy in Poland. [...] [Twardowski] set out to reform [Polish philosophy] by implementing the Warsaw positivists' concept of "organic work", but implementing not the positivist idea, but the method. (Woleński 1985: 13–14)

Finally, then, the concept of philosophy developed and promoted by[4] both the founder of the Lvov-Warsaw School and most of his students is referred to as the scientistic concept (Woleński 1985: 72). Leszek Kołakowski points to the significant role of the Lvov-Warsaw School in the history of modern Polish positivism (Kołakowski 1966: 209–211). Woleński (1985: 26) explicitly opposes such references:

[4]Let us note right away that 'in the [Lvov-Warsaw] School, no special manifestos were written expressing metaphilosophical views—instead, the nature and methods of philosophy were explained on the margin of works concerning particular philosophical problems, and only occasionally in dedicated articles' (Woleński 1985: 52).

'Let me make it clear right away that it is my intention to demonstrate that the Lvov-Warsaw School is, despite many similarities and interdependencies, an entirely distinct philosophical formation, rather than a "Polish neo-positivism"'. Its detailed discussion can be found in Woleński's monograph (Woleński 1985: 35–40, 52–76). Summing up this thread of his reflections, this scholar in the thought of the Lvov-Warsaw School identifies several features characteristic of the way philosophy was practiced there.

The first one is a so-called collective understanding of philosophy. It refers to the fact that thinking in terms of a single 'philosophy' was avoided, and instead it was emphasised that there are many philosophical sciences including various disciplines. The number of these disciplines depends, among other factors, on the stage of their development (Woleński 1985: 37).

Another feature is revealed in the School's preferred approach of solving specific problems rather than that of building 'great philosophical systems', particularly speculative metaphysical systems whose scientific character was criticised. Already Twardowski included these systems in the worldview, which must be embraced by everyone, including a philosopher. He stressed, however, that the worldview must be separated from strictly philosophical reflections (Woleński 1985: 38–39). Yet another feature consists in that when engaging in philosophical deliberations, one must maintain logical accuracy and methodological consistency in the notions, terms and arguments used. Moreover, the relationship between philosophy and language was distinctly emphasised, in particular with respect to language conceived as the subject-matter of philosophical reflection (Woleński 1985: 73).

All of these features eventually determined the understanding of philosophy as a scientific discipline, which was to result in practicing a scientific philosophy. It should be stressed that, unlike in Western European circles related to logical empiricism, the Lvov-Warsaw School did not negate the existence or legitimacy of at least some of the issues traditionally forming part of metaphysics or ontology.

It seems that a similar view of philosophy was also held by Zawirski. Woleński stresses the consistency of the Lvov-Warsaw School in this regard:

> There are in the view of all (or almost all) of the representatives of the Lvov-Warsaw School mentioned above [including Zawirski] elements which fall under its [philosophy's] primary matrix. And without any doubt, all of the above scholars shared Twardowski's general attitude to philosophy, and it was that very attitude that acted as the bond uniting philosophers who had different interests and different views on particular issues. (Woleński 1985: 25)

Zawirski presented his metaphilosophical views in a 1938 article, written in French, entitled 'Science et philosophie' (Zawirski 1938), as well as a treatise—published years later based on a manuscript—entitled 'Science and Metaphysics' (Zawirski 1995, 1996). On another occasion, Zawirski introduced himself as a philosopher in the following words:

> A Polish philosopher [...].Deals with philosophical issues in natural sciences and the theory of natural knowledge, as well as natural issues in general; a follower of the programme of scientific philosophy. [...] Rejecting the rigidity of the notional apparatus [...] he is nevertheless of the opinion that whenever we are forced to modify our principal notions and rules, such changes should be accomplished so that the old notions and rules are the borderline cases of the new ones. [...] The history of mathematical and natural sciences proves the legitimacy of this standpoint. (Jadacki 1998: 203)

And this, in turn, confirms the suggestion made at the beginning of this paragraph. Zawirski believed in particular that philosophy should only make use of methods developed within the field of empirical sciences, which ultimately means deduction and induction. Non-scientific methods, on the other hand, included eidetic cognition proper to phenomenology, as well as intuition as seen in Bergson's views, and neo-Hegelian dialectics. Discussing Zawirski's metaphilosophical views, Michał Heller and Janusz Mączka point out that his objections were the least serious about phenomenology, as 'the eidetic cognition it embraces is at least a type of intellectual cognition, while the other two directions refer to some non-intellectual intuitions' (Heller and Mączka 2004: 235).

For Zawirski, the postulate of using methods proper to empirical sciences in philosophy did not mean doing away with metaphysics as a whole. In this regard, the Lvov-Warsaw School held views which were very different from those of the then popular neo-positivism. Representatives of the latter believed that problems traditionally analysed by philosophy were pointless due to their being unverifiable by definition. For Zawirski, the process of rescuing philosophy, in particular metaphysics, progressed in stages. We are reconstructing it here based on analyses performed by Heller and Mączka (2004: 235–236). First, he pointed out that an important element of the scientific method is deduction. 'Advanced sciences are hardly ever about verifying isolated empirical (protocol) statements, but about their combined wholes. And such a whole is made up of deductive relationships. The meaning of an expression found in such a whole is determined by axioms of the employed system applied through interpretation'. Then, referring to a claim proven by Kurt Gödel, who demonstrated that in axiomatic (deductive) systems containing arithmetic there are true (meaningful) sentences which are not provable within the systems themselves—Zawirski pointed out that '[i]f there are meaningful sentences even in deductive systems which cannot be verified, then the same cannot be ruled out in empirical sciences'. In empirical sciences, in turn, discussions recounted by Zawirski in his article 'Science et philosophie', concerning problems of verification, falsification and reduction of scientific theories by empirical data, allowed him to finally make the final conclusion that even in empirical sciences verifiability cannot be equated with meaningfulness. 'And if meaningfulness does not coincide with empirical verifiability, then there is room for philosophy, and in particular for its most important department – metaphysics. The fact that there has been little development in metaphysics since Ancient Greek times may prove that in this department of philosophy there are issues with are unresolvable and yet meaningful'. Consequently, 'the neo-positivist abolition of philosophy is neutralised'. Moreover, Zawirski believed that there are problems in traditional metaphysics which may be resolved using the scientific method. To illustrate this claim, he mentions issues related to time and space, or disputes between determinism and indeterminism.[5]

[5]'Zawirski became interested in a range of philosophical and natural issues. It will not be an overstatement to say that he considered all major problems presented by the development of physics [e.g. the formulation and development of the theory of relativity and quantum mechanics]' (Heller and Mączka 2004: 237).

Based on the above analyses, it may be concluded that Zawirski was an advocate of reductionism in his metaphilosophical views, but only in its methodological (concerning methods employed to arrive at assertions) and epistemological version (concerning the sources of true knowledge). He objected, however, against reducing ontological philosophy to empirical sciences (that is, against doing away entirely with problems traditionally considered as proper to philosophical discourse). Nevertheless, metaphysics based on experience and the methods of empirical sciences no longer acts as an unchallenged and ultimate system of knowledge about reality. Instead, it should try to combine the results of empirical sciences into a system that is free from contradictions, and 'clearly realise the incomplete and provisional nature of its assertions'. If this was achieved, Zawirski believed, 'it would not matter if we called it "science" or "philosophy"' (Zawirski 1995: 135). Nevertheless, both are based on the same assumption—that there exists some kind of supra-individual reality which is 'more than a sum of the experiences of individual persons' (Zawirski 1996: 136).

To conclude this part of the chapter, we would like to refer to Zawirski's reflections on time. His primary paper on the subject was written in French and entitled 'L'évolution de la notion du temps', published in 1936 under the auspices of the Polish Academy of Arts and Sciences (Zawirski 1936a). In the same year, Zawirski also published an abridged text of the same paper in 'Kwartalnik Filozoficzny' (Zawirski 1936b), to which we will refer in analysing his approach to philosophical reflections on time.

Discussing the evolution of the notion of time, he lists the concepts proposed by Aristotle, Kant or Bergson, as well as relativist physics and quantum mechanics 'in the same breath' (Zawirski 1936b: 48). He does not talk about any 'shift' from one type of knowledge (philosophical) to the other (scientific). Moreover, he emphasises that thanks to the nineteenth-century criticism of Kant's concept of time, and above all Newton's theory of time, the door was opened to twentieth-century proposals in the form of theories developed by Bergson, Husserl, and first of all—Zawirski stresses—the theory of relativity. In his description of twentieth-century views on time, he also lists the concepts proposed by Poincaré, Enriques, and quantum mechanics (Zawirski 1936b: 49).

We can thus clearly see that Zawirski—with regard to the evolution of the notion of time—treated the achievements of science and philosophy in a synthetic way. Knowledge about time should be cumulative, combining the achievements of both science and philosophy. This, naturally, is consistent with his metaphilosophical views on the nature of knowledge developed in scientific philosophy.

Two main groups of issues related to time mentioned in Zawirski's philosophical reflections concern the problem of the reality of time and its genesis (Zawirski 1936b: 50). With regard to the former, Zawirski referred to three standpoints. Next to extreme realism and idealism, he also distinguished moderate realism. Extreme realism said that time is independent in its existence from any other beings or phenomena. This meant that, according to the advocates of this standpoint, time was an autonomous substance. A typical representative of extreme realism in the understanding of time was Newton. Idealists, on the other hand, negated the real existence of time. Zawirski points out that such idealism can be found in the views of Immanuel Kant. 'For Kant, in turn, the temporal aspect of the world would disappear with the disappearance of consciousness, or at least if consciousness were deprived of its form of capturing phenomena' (Zawirski 1936b: 51). This is what both time and space represented for this thinker. The world 'in itself', on the other hand, is cognitively inaccessible, so according to Kant it is not possible to determine whether it is a temporal being, or whether it exists outside of time. The third standpoint on the issue of the existential status of time is moderate realism. According to Zawirski, it is most strongly represented in the history of philosophy, for moderate realists included both Aristotle, the scholastics, Leibniz, and Bergson. Zawirski adds that 'extreme realism, just like idealism, was only sporadically encountered in both antiquity and the modern times' (Zawirski 1936b: 51). Moderate realism says that while time is not an autonomous substance, neither is it a thing (in this respect, Zawirski agrees with the views of Tadeusz Kotarbiński for whom time was a so-called apparent name), yet it is something real, something that coexists with real things. Time may either be a property of really existing things, or a feature of such properties. Zawirski emphasises that 'for a moderate realist, time would only disappear with the disappearance of the world or its transformations' (Zawirski 1936b: 51).

As regards the genesis of the notion of time, Zawirski points to its two main sources. The first are 'facts of spiritual life', and the other 'data of the physical world'. In particular standpoints belonging to either of the two forms of realism or to idealism, when answering the question about the genesis of the notion of time, these sources may be treated separately, i.e. we may either speak only of facts of spiritual life or only of data of the physical world, or both of them can be seen together, treated synthetically (Zawirski 1936b: 52). The former type of theories of the genesis of time, in Zawirski's nomenclature referred to as 'psychological theories', emphasise the significance of qualitative properties of time. The latter theories, which Zawirski identifies with natural theories, emphasise the quantitative properties of time, identified with its metric properties. He concludes his discussion of the subject with a comment that is significant in terms of our reflections here, which is why we will cite it here at length:

> The opposition of these two directions can also be observed throughout the history of philosophy and sciences. The Platonists, Saint Augustine, Locke are, as though, stages in the evolution which has led to Bergson. The orientation of Aristotle and Thomas Aquinas is related to modern physics, on the other hand. It goes without saying that both of these tendencies should be reconciled, since time proves to have both quantitative and qualitative properties. (Zawirski 1936b: 52)

We would like to reflect for a while on these words, because they illustrate how Zawirski applies his metaphilosophical postulates in practice. First of all, it should be noted that with respect to studies on time, Zawirski believes that both the scientific perspective and the meta-empirical dimension, going beyond data provided by natural sciences, should be taken into account (Zawirski 1996: 133). Only this way is a 'complete' ontological description of time possible, i.e. one that takes into account both its qualitative and quantitative properties. The question that may now be asked is whether Zawirski consistently applies the postulates of scientific philosophy to deliberations on time. It appears that in order to fully understand his standpoint, we should refer to a passage from 'Science and Metaphysics' (Zawirski 1996: 134–135). On the one hand, Zawirski identifies the point where the roads of 'science' and 'philosophy' (or rather 'science' and 'metaphysics'), so far undistinguishable from one another, become separated:

> Clearly, we intuitively embrace the existence of a supra-individual reality, for instance the world in space; we accept that everything that exists in temporal and special dimensions is a being; a metaphysician, when talking about being, means *ens realissimum*, *ens perfectissimum*, the only, eternal Being. We can find the laws of logic, but we cannot find the absolute Being inside ourselves in the same way. (Zawirski 1996: 134)

On the other hand, however, also 'the world studied by science is not given to anyone directly. For every person experiences their own world; we can say that there are as many worlds as there are living creatures' (Zawirski 1996: 134). Zawirski points out that:

> [T]he main contribution of modern philosophy, particularly English idealism, and in part also of Descartes, is pointing to the fact that belief in the existence of a physical world, in any form whatsoever, goes beyond the data of experience; it is the same kind of transcendence which we embrace when we accept [the existence of] God outside of the world. (Zawirski 1996: 134)

It thus turns out that the moment of transcendence is common to both science and metaphysics.

Going back to the main thread of these reflections, to sum up the analyses performed so far, it should be said that in the context of Zawirski's views cited above, a combination of the strictly scientific approach, scientific philosophy, and some metaphysical content is possible, and with respect to reflections on time even seems to be necessary if we want to adequately depict the entire problem of the phenomenon of time, of which already Saint Augustine wrote in Book XI of his 'Confessions': 'What, then, is time? If no one asks me, I know what it is. If I wish to explain it to him who asks me, I do not know' (Augustine 1955: ch. XIV, 17).

This way, we can see a clear difference between Zawirski's metaphilosophical views and the postulates that all sciences be reduced to physics, and all metaphysical (meta-empirical) reflections be removed outside of rational discourse.

3 Multidimensionality of the Conceptual Framework Related to the Postulate of Interdisciplinarity

The purpose of analyses we have performed so far was to prepare grounds on the basis of which we will now try to evaluate whether Zawirski's philosophical analyses concerning time and his attitude resulting from the scientific method of philosophy he embraced may have some connotations with contemporary postulates concerning the interdisciplinarity of scientific research.

Before we do this, however, we must first make one more comment. In order to say whether the method employed by Zawirski in his philosophical reflections and his metaphilosophical views fall within any of the concepts related to the problem of interdisciplinarity, i.e. the concepts of inter-, multi-, or trans-disciplinarity, it is first necessary to at least briefly specify what these concepts refer to. To this end, we will refer to analyses, already cited above, presented by Poczobut in his article entitled 'Interdyscyplinarność i pojęcia pokrewne' ('Interdisciplinarity and Related Concepts').

First, we would like to focus on the concept of multidisciplinarity. When referring to multidiscipline, Poczobut has in mind a free-floating coalition of various scientific disciplines, which are not merely a simple collection of practically any elements (disciplines), however. 'There needs to be some kind of integration among them, a shared range of problems (even if only roughly sketched out), as well as some institutional setting within the structure of science' (Poczobut 2012: 44). Multidisciplinary studies result from the observation that for many research problems, particularly those with multiple dimensions, or which encompass many different aspects, a methodological isolationism and a mono-disciplinary approach are not sufficient. Naturally, every multi-discipline consists of sub-disciplines, which may also be of diverse nature. This means that a multidiscipline may include both monodisciplines and other formations, such as interdisciplines or even more abstract (as we will try to demonstrate shortly) transdisciplines. The degree of integration within a given multidiscipline depends on this content. In the context of his analyses, Poczobut points out that in principle 'every sufficiently developed discipline of science is in fact a multidiscipline made up of many specialised sub-disciplines' (Poczobut 2012: 47). Examples of multidiscipline include both traditional disciplines of science such as mathematics, physics, or biology, and philosophy. 'In spite of the fact that it includes such distant disciplines as the philosophy of mathematics and aesthetics, there is a strong conviction (rarely explicated, but supported by institutional factors) that they create a single multidisciplinary field' (Poczobut 2012: 47).

Another concept we would like to briefly discuss is the notion of interdisciplinarity mentioned in the title. The emphasis here is on the decidedly greater—than in the concept of multidisciplinarity—interdependence between disciplines making up an interdiscipline. Interdisciplinary studies attempt to solve problems existing on the interfaces of the primary disciplines. 'Interdisciplinarity always involves identification of explanatory boundaries irrespective of the existing disciplines and theories' (Poczobut 2012: 49).

To put it simply, there are problems a correct solution to which can only be proposed in the framework of studies encompassing more than one discipline and requiring the cooperation of representatives of various disciplines of science, e.g. physicists with chemists, chemists with biologists, biologists with social scientists, etc. Poczobut, quoting Mario Bunge, identifies conditions necessary for a new interdiscipline to emerge. First, there must exist at least two well-defined independent disciplines of science (underlying disciplines) on the interface of which a new interdiscipline will be created. Secondly, there must be some common area shared by the underlying disciplines (i.e. they should be concerned, at least in part, with the same objects). Thirdly, there must exist (be constructed) so-called 'glue formulas' (Bunge 2003: 278–280, 286) containing concepts and terms proper to each of the underlying disciplines. 'Their role is to bind the languages and conceptual systems of the theories (disciplines) [...] they represent relationships between their disciplines' (Poczobut 2012: 54). If this results in the creation of a new interdiscipline, then in the next stage it is bound by the same requirements which apply to traditional monodisciplines:

> The important thing is that in each case the formulas and binding languages (conceptual systems) of the various disciplines concerned should be submitted to the same empirical tests as hypotheses formulated on the grounds of standard monodisciplines. Some of them may be subject to falsification, other may successfully pass empirical tests, thus contributing to the establishment of a given interdiscipline as a science. The ultimate instance which all interdisciplines must refer to are the empirically available interdependencies between the fields of their underlying disciplines. (Poczobut 2012: 55)

In relation to the concept of interdisciplinarity, Poczobut refers to Mario Bunge's more general thought (Bunge 2003: 281–283): 'The entire system of knowledge resembles a rosette of hundreds of partially overlapping ellipses representing individual disciplines of science. In this image, the intersecting areas (common parts) symbolise interdisciplines and interdisciplinary research programmes' (Poczobut 2012: 55). Moreover, Bunge, cited by Poczobut, concludes that in our system of knowledge:

> [T]here are no entirely independent (absolutely isolated) disciplines. If the field of each discipline of science partially overlaps with the fields of study of other disciplines, then—strictly speaking—there are no radically monodisciplinary fields of study. Moreover, if a discipline does not have any (substantial and conceptual) relationships with other disciplines, then it is not part of the comprehensive system of knowledge (it is not a science). (Poczobut 2012: 56)

Consequently on this basis Bunge (and Poczobut) suggests the introduction of a metascientific norm to co-define, next to other norms of the kind, such as internal consistency or empirical testability, the content of the notion of science. This norm is to be a so-called condition of external coherence:

> External coherence does not consist in a banal compatibility of studies, which we are dealing with when two areas of study (two disciplines or theories) have no common referents, and for which no conceptual inter-theoretical and interdisciplinary relationships exist. Rather, it is about consistency with the statements and theories of those disciplines which the one concerned remains in interdisciplinary relationships with. (Poczobut 2012: 56)

The goal is to counteract the unwillingness to confront the results of one's own studies (statements and theories) with well-tested data coming from related disciplines. Interestingly, in addition to examples from natural sciences (chemistry, biology), Poczobut also talks about philosophy: 'a philosopher of the mind, when building his theories, should see if his results are consistent with the results of relevant empirical sciences' (Poczobut 2012: 56).

Finally, the last term, transdisciplinarity, refers to studies which go beyond the standard disciplines of science. Poczobut illustrates the essence of transdisciplinarity by confronting it with interdisciplinarity:

> Unlike interdisciplinary studies aimed at capturing properties common to diverse implementations of general categories in the discourse of individual disciplines of science, transdisciplinary studies are conducted on a higher level of abstraction. […] Transdisciplinarity enters the stage when using various levels of representing reality, as well as all knowledge available within mono- and interdisciplinary studies, we construct theories on an even higher level of abstraction. (Poczobut 2012: 58)

Interdisciplinary research is to be focused on the implementations of a particular general category in a given field of science, while transdisciplinary research—on a general category conceived independently from any particular exemplification. It is clear right away—as Poczobut also points out—that transdisciplinary studies described this way have a lot in common with analyses carried out on the grounds of ontology or epistemology. 'In the case of basic ontological and epistemological categories, the general outline is the same—we have categories characterised in abstract terms and their field applications, which attests to the transdisciplinary nature of ontological and epistemological studies, as well as the categories they are concerned with' (Poczobut 2012: 59).

4 Conclusion

Having presented key notions related to the idea of interdisciplinarity in science, we can now attempt to find links, if any, between them and the metaphilosophical views of Zawirski (or, more broadly, at least some of the views characteristic of the Lvov-Warsaw School as a whole). What is most apparent is an affinity between the collective understanding of philosophy declared by the Lvov-Warsaw School and the concept of multidisciplinarity. In this regard, there seem to be no major differences between them.

We would therefore venture to propose that the postulate of multi-discipinarity, or of treating philosophy as a multidiscipline, was actually implemented in the Lvov-Warsaw School. In particular in personal terms, when individual scholars investigated into specific problems moving within a certain multidisciplinary field. In the case of Zawirski, this was expressed in studying multidimensional (multifaceted) problems, such as the issue of time, causality, or determinism of natural laws.

These analyses also show, particularly as regards problems related to time, that in his studies Zawirski moved within the area of interdisciplinarity. His analyses of the evolution of the concept of time clearly show that with this problem he believed it was necessary to develop something which—to use today's terminology—we could consider an interdisciplinary research programme. This is especially true about the postulate that qualitative and quantitative properties of time should be combined. The only difference, though a rather significant one, compared to contemporary calls for interdisciplinary research, consists in that in the background of Zawirski's reflections there seems to be an intention to—as far as possible—integrate sciences. This seems to reveal the prevalence of the thesis about the unity of science in Zawirski's times, one of the foundations of the view related to the scientistic approach. Today, no one seems to seriously think that all disciplines could ever be united under the aegis of a single 'superdiscipline'. It also appears to me that Zawirski would not agree with the image of the system of knowledge as a rosette in which it is virtually impossible to determine the borders between particular fields of knowledge. On the other hand, his reflections included a postulate, appearing also in the context of multidisciplinary studies, concerning the standard of external coherence of individual disciplines of science.

As for the concept of transdisciplinarity, we believe also in this matter Zawirski might have some objections. At the time when he lived and worked, a general theory of systems had not been built yet, and the construction of abstract systems from general notions would resemble too much purely speculative metaphysical systems, entirely detached from the world of sensual experience. Nevertheless, from today's perspective, knowledge about the world forming a hierarchically structured system which includes both mono-disciplines and a great variety of multi-, inter- and transdisciplines seems to be very useful and helpful in building a rational and adequate picture of the world.

References

Augustine, St. (1955). *Confessions* (A. C. Outler, Trans.). https://www.ling.upenn.edu/courses/hum100/augustinconf.pdf.

Bunge, M. (2003). *Emergence and convergence: Qualitative novelty and the unity of knowledge*. Toronto, Buffalo, and London: University of Toronto Press.

Chmielewski, A., Dudzikowa, M., & Grobler, A. (Eds.). (2012). *Interdyscyplinarnie o interdyscyplinarności. Między ideą a praktyką*. Warszawa: Wydawnictwo Impuls.

Duchliński, P. (2014). *W stronę aporetycznej filozofii klasycznej. Konfrontacja tomizmu egzystencjalnego z wybranymi koncepcjami filozofii współczesnej*. Kraków: Akademia Ignatianum, Wydawnictwo WAM.

Fordeman, R. (Ed.). (2010). *The Oxford handbook of interdisciplinarity*. Oxford and New York: Oxford University Press.

Hejmej, A. (2007). Interdyscyplinarność i badania komparatystyczne. *Wielogłos. Pismo Wydziału Polonistyki UJ, 1*(1), 35–53.

Heller, M., & Mączka, J. (2004). Krakowska filozofia przyrody w okresie międzywojennym. *Prace Komisji Historii Nauki Polskiej Akademii Umiejętności, 6*, 213–243.

Jadacki, J. J. (1998). *Orientacja i doktryny filozoficzne. Z dziejów myśli polskiej.* Warszawa: Wydział Filozofii i Socjologii UW.
Klein, J. T. (1990). *Interdisciplinarity: History, theory, and practice.* Detroit: Wayne State University Press.
Kołakowski, L. (1966). *Filozofia pozytywistyczna. Od Hume'a do Koła Wiedeńskiego.* Warszawa: PWN.
Mizińska, J. (2012). 'Człowiek to człowiek'. Esej o kulturowych i światopoglądowych przesłankach problemu interdyscyplinarności. In A. Chmielewski, M. Dudzikowa, & A. Grobler (Eds.), *Interdyscyplinarnie o interdyscyplinarności. Między ideą a praktyką* (pp. 71–84). Warszawa: Wydawnictwo Impuls.
Poczobut, R. (2012). Interdyscyplinarność i pojęcia pokrewne. In A. Chmielewski, M. Dudzikowa, & A. Grobler (Eds.), *Interdyscyplinarnie o interdyscyplinarności. Między ideą a praktyką* (pp. 39–61). Warszawa: Wydawnictwo Impuls.
Turek, J. (2009). *Filozoficzne interpretacje faktów naukowych.* Lublin: Wydawnictwo KUL.
Twardowski, K. (1904). Przemówienie na otwarciu Polskiego Towarzystwa Filozoficznego we Lwowie. *Przegląd Filozoficzny, 7*(2), 239–243.
Walczak, M. (2016). Czy możliwa jest wiedza interdyscyplinarna? *Zagadnienia naukoznawstwa, 207,* 113–126.
Woleński, J. (1985). *Filozoficzna Szkoła Lwowsko-Warszawska.* Warszawa: PWN.
Zawirski, Z. (1936a). *L'évolution de la notiondutemps.* Kraków: Académie Polonaise des Sciences et des Artes, Librairie Gebethner et Wolf.
Zawirski, Z. (1936b). Rozwój pojęcia czasu. *Kwartalnik Filozoficzny, 12*(1), 48–121.
Zawirski, Z. (1938). Science et philosophie. *Organon, 2,* 1–16.
Zawirski, Z. (1995). Nauka i metafizyka: cz. 1. *Filozofia nauki, 3, 3*(11), 103–135.
Zawirski, Z. (1996). Nauka i metafizyka: cz. 2. *Filozofia nauki, 4, 1*(13), 131–143.

14

The Informational Worldview and Conceptual Apparatus

Paweł Stacewicz

1 Introduction

This chapter relates to the thought of the Lvov-Warsaw School in a most contemporary way, as it refers to computer science—a domain which did not exist yet in the main period of the School's activity.

Such reference is substantiated by two reasons. Firstly, because both Kazimierz Twardowski and most of his students viewed philosophy as a discipline which should develop in interaction with other scientific disciplines (which today include computer science). Secondly, because one of the prominent contemporary continuators of the School, Professor Witold Marciszewski, has used concepts related to computer science—along with their logical and mathematical background—to clarify various philosophical issues. This paper is anchored in this current of research, and its final section is devoted to the discussion of a very

P. Stacewicz (✉)
Faculty of Administration and Social Sciences,
Warsaw University of Technology, Warsaw, Poland
e-mail: p.stacewicz@ans.pw.edu.pl

A. Drabarek et al. (eds.), *Interdisciplinary Investigations into the Lvov-Warsaw School*, History of Analytic Philosophy,
https://doi.org/10.1007/978-3-030-24486-6_14

particular concept proposed by Marciszewski, namely that of 'cognitive optimism'.

An additional reason for including this essay in this volume is the fact that it presents a certain type of exploration into worldviews. And these, in accordance with declarations made by the School's representatives quoted further on, have always been (and should always be) an important driving force behind philosophical investigations.

2 Between Worldview and Philosophy

The notion of worldview, one of key importance for this text, is closely related to philosophy, as the sources of philosophical investigations are most explicitly rooted in the worldview needs of individual people and the societies they make up.

In the individual (and psychological) dimension, a worldview may be defined as a collection of one's fundamental beliefs about issues of such existential importance as the structure and knowability of the world, ethical and moral values, the existence of God, or the nature of truth. In the practice of everyday life, such beliefs act as a guidepost, determining one's goals and pursuits.[1] In this sense, they are a kind of small, private philosophies, expressing one's attitude towards the world.

In the social dimension, worldview is even more closely related to philosophy, as it represents an intersubjective phenomenon (one that goes beyond individual experience and activity), sufficiently well-established in the awareness of many people, having its preliminary expression in works of art, science, or even economic systems characteristic of a particular age. In this sense, we can talk of Christian, romantic, rationalist, mechanicist, informational, or other worldviews. As will be explained later on, such socially established types of views may both

---

[1]Marciszewski uses a metaphorical expression here: 'a compass guiding one's behaviour' (Marciszewski and Stacewicz 2011: 223). On the notion of worldview (cf. Marciszewski 1996: 133–136).

contribute to the development of certain philosophical directions, and be inspired by some philosophical systems.

There is one more issue to be touched upon here. A scientific concept of worldview understood in its social aspect must be abstract—i.e. independent of the experiences and beliefs of individual people. In such approach, the point is to describe a certain idealized type of views which is, however, not only scientifically valid (i.e. yields to abstract analyses and comparison with other types of views), but may also really influence—as a notional matrix—the worldviews of individual people.

The relationship between philosophy and explorations into worldview appears to be entirely natural, as has been repeatedly demonstrated by philosophers who wrote about the specific nature and functions of their discipline. Due to the profile of this paper, we will limit our discussion to a selection of comments by representatives of the Lvov-Warsaw School.

The first one to mention is Władysław Tatarkiewicz, who says in the introduction to his 'History of Philosophy':

> For as long as science has existed, efforts have been made to go beyond particular considerations, and include all that exists in single science; attempts have never been given up at building, along with particular sciences, one that would provide a *view of the world*: this science has been and still is called philosophy. (Tatarkiewicz 1990: 5)

The above remark corresponds to the following observation made by the School's founder, Twardowski, in his famous address delivered on the occasion of the 25th anniversary of the Polish Philosophical Society:

> Indeed, one might say that it is the most important aspect of any properly conceived philosophical work, as the development of just such a *view* (i.e. a view on life and the world - P.S.) has been the goal of the most sublime creative efforts of the greatest philosophers of all times. (Twardowski 1965: 379)

The involvement of philosophy in the issues of worldview was also strongly emphasized by Kazimierz Ajdukiewicz—who saw metaphysics and the epistemology that supports it as the 'road' to the ultimate, most comprehensive view of the world (Ajdukiewicz 2003: 143–150).

The views cited above, expressed by very influential representatives of Polish philosophy, illustrate well this dimension of the relationship [*worldview-philosophy*], expressed in the desired and expected influence of philosophy (its mature systems) on the worldviews of individual people.

This relationship also has another dimension, however, revealing the opposite direction: this time moving 'from worldview to philosophy'. This is explained in the following argumentation by Marciszewski:

> The boundary between worldview and philosophy is not a sharp one, yet some of it can be delineated. We talk of worldview when we have in mind a view of the world which develops in a community in a rather spontaneous way, as a result of circumstances and demands of the time; philosophy, on the other hand, is *a more systematic and theoretical development of these thoughts* by specific authors. As an example, we can look at the history of Christianity. The Christian worldview is already present in the gospels and letters of the apostles as a religious answer to the contemporary question about the meaning and purpose of life. Christian philosophy, on the other hand, does not appear even in its embryonic form until a whole century later, for example in Origen… (Marciszewski and Stacewicz 2011: 219)

What, then, is the relationship between worldview and philosophy? Since it is reciprocal and consists in mutual influences, it may be referred to as that of feedback occurring over time. As the term suggests, philosophy develops due to spontaneously emerging worldviews which are, on the other hand, shaped by this very same philosophy through its sufficiently well-developed systems.

This description would not be complete without a reference to the relationship between worldview and philosophy on the one hand, and specific sciences (including empirical ones) on the other. These sciences interact with the worldview and philosophy, thus forming yet another—we might say: external and superior to the two—level of feedback. What occurs at this new level is, firstly, that philosophies related to worldviews become scientified, and secondly, that certain philosophical and worldview intuitions (such as the Platonic intuition about ideas, for instance) penetrate and reinvigorate the sciences.

In this context, let us quote yet another comment by Twardowski:

> By providing a scientific analysis of initially metaphysical views, individual sciences cooperate in building a scientific view of the world and life, and since the same scientific view is also pursued by the authors of metaphysical views themselves, as long as they reckon with the results of particular scientific studies, a relationship of reciprocity is formed: particular sciences draw on some ideas, notions and theses developed in metaphysical systems, and metaphysical systems receive these ideas, notions and theses back from those sciences in a scientified form. (Twardowski 1965: 383)[2]

Within this framework of interactions, we might say that in some ages certain sciences respond particularly well to the current challenges of civilization and exert the greatest influence on social worldviews (thus contributing to the development of new types of philosophies). The informational worldview analysed further on in this article is, in my opinion, just such a contemporary expression of the significant social role of a particular scientific discipline, namely that of computer science.

Being a collection of philosophically significant beliefs, a worldview requires, naturally, a set of notions to express these beliefs. Importantly, however, these notions—properly selected and arranged in a hierarchy—are not only an indispensable means of expression, but also a preliminarily condition for the so-called picture of the world that underlies a worldview. Referring to the terminology once used by Ajdukiewicz, we might say that they delineate a certain perspective on the world, which here, as we are considering worldview issues, we will

[2]It should be stressed that Twardowski treated the scientific worldview (or: the ultimately scientifically substantiated view of the world and life) as an ideal to be pursued, but never to be actually achieved.

refer to as a worldview perspective.[3] For example, in the mechanicist perspective, typical of the eighteenth century scientific community, the concepts that moved to the foreground were those of motion and machine; in the informational picture we are interested in here, the leading ideas are algorithm, code, or computability.

To emphasize even more the relationship between a worldview and its underlying picture of the world, the latter may be referred to as an outlook on the world. As we have already mentioned above, to arrive at an outlook on the world we must use a set of notions, which apparently plays two roles. Firstly, it enables us to narrow down our field of vision (perspective) to that which is preliminarily considered relevant; and secondly, it allows us to add focus and depth to the picture (e.g. providing insight into the hidden structure of material objects). Importantly, however, these two functions do not lead directly to judgments constituting a worldview. For such judgments to be formulated, it is necessary to adopt a stance on the world, and not only to choose a point of view (or: a notional framework constituting the worldview perspective).

The intermediate link which allows us to move gradually from an outlook on the world to a 'stronger' worldview are questions. On the one hand, they may only be asked on the condition that we adopt a particular perspective (outlook), and on the other—the way they are answered leads to a certain set of beliefs about the world (worldview).

It is now time to illustrate the above distinctions with suggestive examples, which will at the same time be an introduction into a more systematic discussion of the informational worldview.

[3]The term 'world perspective' was used by Ajdukiewicz in the context of the methodology of sciences, referring to a set of beliefs (or, more properly, statements) characteristic of a particular science at a particular stage in its development. These judgments make up the knowledge proper to a particular science about a particular fragment and aspect of the world. Thus, every science develops its special perspective on the world (rather than a comprehensive picture of the world), consistent with the findings it has made so far. Such perspective may be treated (as we do in this text) as a way of viewing the world offered by a particular science as a framework within which one can build a worldview going beyond science—i.e. a set of views which are not unequivocally ruled out by scientific findings. In this new context, a more adequate term is therefore 'worldview perspective' (rather than 'world perspective') (cf. Ajdukiewicz 1934: 409–416).

In today's society—more and more often now referred to as information society or computerized society[4]—there is an ever stronger tendency to describe a growing number of phenomena in informational categories. For instance, the human mind is compared to an information-processing system (e.g. in cognitive science); the abilities and development of living organisms are explained using properties of the DNA code (which is a physical form of recorded information); sometimes even the entire universe is compared to a gigantic computer (e.g. by physicists such as Ed Fredkin or Seth Lloyd). All such operations introduce into our culture, to an ever increasing extent, an informational outlook on the world, or an informationally founded way of describing reality. Notions which play a key role in its structure include: information and data, algorithm and programme, computability and uncomputability.

The world categorized using the above-mentioned notions may be the object of various questions, the—even provisional—answers to which make up the informational worldview (rather than just an outlook). Let us provide two examples:

1. Is the physical world around us analogue, digital, or both (in the sense best defined in the field of computer science)?
2. Is the computational power of the human mind greater than that of digital computers (theoretically described using Turing machines)?

Let us notice that such questions could not even be put into words if the world were not described using appropriately interpreted computational categories.

[4]The concept of information society was introduced by a Japanese, Tadao Umesao (in 1963), and then popularized by the American sociologist Daniel Bell (beginning in the 1980s). It describes a society in which: (a) the dominating sector of economy is services (including information services); (b) the economy is based on knowledge; (c) the occupational structure is dominated by specialists and scientists; (d) there is a strong development of new (information and communication) technologies of managing information and knowledge; and (e) there is an increasing tendency to computerize more and more areas (Ito 1991: 3–12).

3 The Informational Conceptual Apparatus[5]

The conceptual apparatus which is proper to computer science and which determines the informational worldview perspective is a formal one.[6] This means that basic concepts used in computer science, such as algorithm, data structure, or computational complexity, are defined and then examined within the framework of abstract formal theories (such as the theory of computation). Moreover, and more importantly for us, they may be provided with various specific interpretations, including technical, cognitivistic, economic or philosophical ones. For example, asking 'Is the computational power of the human mind greater than that of a universal Turing machine?', we refer to cognitive science (and the related interpretation). Within its framework, a mathematically defined machine (mathematical object) is referred to human cognitive

[5]The title of this section clearly refers to the term 'conceptual apparatus' used by Ajdukiewicz. This reference is substantiated, as in the context of particular sciences Ajdukiewicz defined conceptual apparatus as the set of meanings of expressions (both terms and sentences) formulated in the language of that science, which make it imperative—sometimes in the presence of empirical data, and sometimes without them—to accept some judgments and reject others. Thus, they delineate the picture (or description) of the world proper to the science concerned at a given stage in its development (Ajdukiewicz 2006: 175–195).

The informational conceptual apparatus defines a picture characteristic of computer science. It thus represents a certain temporary set of the meanings of terms used in computer science, understood as formal expressions (which may, however, be provided with worldview-relevant interpretations).

[6]By way of a longer footnote, let us explain that the methodological status of computer science is not unequivocal. Depending on how its main concepts (such as algorithm or data) are interpreted and used, computer science is considered to be: (a) a formal science—akin to mathematics; (b) an engineering and technical science—akin to electronics; or even (c) a natural science—in some respects akin to traditional natural sciences (such as physics or biology; in this context, the term natural computing is used). It should be borne in mind that these descriptions are not mutually exclusive; rather, they suggest various aspects of research and applications, instead of defining computer science as such (Knuth 1974: 323–343; Denning 2005: 27–31; Murawski 2014).

Despite the multifaceted nature of studies mentioned above, the central feature of computer science—the one which enables both engineering applications and references to nature—is its formalism. This means that the objects of computer science are first of all formal objects, and only then, within the framework of particular implementations and applications, do they become physical objects (e.g. signals corresponding to particular types of data), or technical ones (e.g. specific, appropriately constructed computer systems). Or, in other words, in order for any products of applied computer science to take form, a theory is required which is developed in a formal (quasi-mathematical) way within such disciplines as algorithmics or computational theory.

activity, and its abstract operations are given the status of operations performed by the human mind.[7]

Thus, the informational conceptual apparatus has two layers: (a) a formal one—describing the most general meaning of particular terms and their formal consequences (e.g. the concept of a universal Turing machine implies that for all instances of such a machine, including physical ones, there exist some uncomputable problems) and (b) an interpretative one—defining certain ways of applying concepts of theoretical computer science to other domains (e.g. when comparing the mind to a Turing machine, additional limitations must be imposed on the latter related to the empirically observable properties of cognitive activity).

It should be stressed that both of these layers are essential for philosophy and worldview alike. Metaphorically speaking, the former opens up computer science to various philosophical issues (permitting strictly philosophical interpretations of computer science objects; interpreting algorithm as an ontological form is a good example). The latter, in turn, expands philosophy's terms of reference to a world whose various fragments and aspects are increasingly described, irrespective of philosophy, in the language of computer science. In the latter case, philosophy draws on computer science in an indirect way. By referring to the findings of another science (e.g. cognitive science), it can access knowledge which is formulated using appropriately interpreted concepts of computer science. The usefulness of these concepts certainly enhances worldview beliefs about the informational or computational nature of the phenomena around us.

Not all of the multitude of formal concepts used in theoretical computer science are equally important for the worldview concept presented here. For the purposes of our discussion, we will list just a few of the

[7]Another example of a cognitivist interpretation is that of the artificial neural network examined as part of research into artificial intelligence. While a particular network (e.g. multilayer perceptron) is examined independently from its various possible applications (e.g. in relation to the question about the best mathematically substantiated learning algorithm), it remains a formal object. If such a network is understood as a model of perception (explaining how the human mind recognizes and classifies objects in a particular field), it is endowed with a cognitivist meaning (Żurada 1992).

basic concepts here, and each will be provided with a short, informal description[8]:

a. information—the content of information processing systems, coded in the form of data which can be processed by a particular system (e.g. digital), described mathematically using certain types of numbers (e.g. natural numbers);
b. computation– the processing of data using a particular set of rules, executable by a particular machine (in the mathematical layer, it is the process of executing a given function, e.g. a recursive one);
c. algorithm—a scheme of computations performed on a machine of a particular type (e.g. digital), which can be formally analysed;
d. model of computation—a set of rules defining permitted computations (e.g. in the Turing model, operations on continuous data are not permitted, while any discrete operations on discrete data are permitted);
e. universal Turing machine—an abstract machine whose structure and operating rules define computations permitted in the digital model (Turing 1936);
f. computational complexity of an algorithm[9]—a type of function linking the volume of an algorithm's input data to the number of elementary operations (in a particular computing model) necessary to complete the computations described by that algorithm;
g. computable problem—a problem for which there exists (in a given model of computation) an effective algorithm able to find its solution;
h. uncomputable problem—a problem for which no effective algorithm exists (in a given model of computation) able to find its solution; an example of an uncomputable problem in the Turing model (see item f) is the halting problem (Turing 1936: 246–249);

[8]A more detailed explanation of these and other concepts can be found in (Stacewicz 2016).

[9]Since elementary operations performed by a machine take a particular amount of time, computational complexity (in the sense defined above) is also called time complexity. Other types of computational complexity are identified as well: including that referring to memory (the amount of the machine's memory units that need to be used), and structure (concerning the complexity of the algorithm/programme itself, i.e. the instructions which make it up).

i. computational power of a model (machine)—the range of solvable (computable) problems in a given model of computation.

Within the conceptual apparatus defined above (only a certain basic fragment of which has been described, and in very general terms), various worldviews can be formulated, whose common element consists in attempting to answer important philosophical questions using terminology taken from computer science.

4 Informational Worldview

Let us now recall that we considered questions to be an element bringing us from an outlook on the world (worldview perspective) to a 'stronger' worldview—questions which may be formulated using the informational conceptual apparatus.

In accordance with this approach, we must agree that various combinations of answers to informationally inspired questions lead to various informational worldviews. In other words, each combination corresponds to a different set of views making up a particular worldview. Thus, if questions are asked following a particular pattern (see the two examples at the end of the previous chapter), then each combination of answers will lead to a certain variant form of the informational worldview.

Consequently, an informational worldview in the abstract sense—let us call it IWV—should be defined as the class of all possible worldviews (sets of beliefs) determined by specially selected, informationally inspired questions.[10]

[10]The term used hereinafter, 'informational worldview', first appeared in the book 'Umysł—Komputer—Świat. O zagadce umysłu z informatycznego punktu widzenia'. Since its publications, various aspects of the informationally inspired worldview have been presented and discussed (in Polish) on an academic blog called Cafe Aleph (http://marciszewski.eu/).

Using a single synthetic formula, an informational worldview can be defined as a certain type of pre-philosophical views which are scientifically grounded and which derive from a strong belief that the key role in describing the world and the relationship between man and the world is played by information technology concepts (such as data, code, algorithm, or computability; cf. Stacewicz 2016: 36).

The above description (IWV as a class) is logically substantiated, but has the drawback of being applicable either to a theoretical situation (when we consider hypothetical questions which we do not formulate directly), or to a precisely defined situation (when we know all of the questions that come into play). The latter situation could occur if the informational worldview were a closed, historical phenomenon, described in view of well-known beliefs of the members of a particular society—beliefs which can be ordered with reference to a closed set of questions.

It is not the case, however. The worldview we are discussing is emerging before our very eyes: it is a living phenomenon, one which changes over time, evolves in line with developments in the constantly improved information technology (and its underlying theories).[11]

Consequently, in fact, no questions can be formulated such as would strictly delineate particular varieties of the IWV. So even though from the logical point of view IWV is a class, and an internally diversified one, its ultimate structure cannot be defined. One can, at the most, list some provisional questions which at the present stage in the development of computer science (both theoretical and practical) appear particularly relevant (like the two questions mentioned at the end of the first chapter).

Despite the above objections, there is yet another method of defining the informational worldview in a broad and abstract sense. For we may identify some general theses which should be accepted by anyone whose worldview is a variety of the IWV. Such theses are a kind of preliminary assumptions which make it possible to ask relevant questions, and consequently to embrace a particular version of the IWV. Thus, they are more fundamental (and permanent) than any questions.

Having emphasized once again that the phenomenon which is here referred to as informational worldview is still developing, we would like to provide three theses as an example. Each will be provided with a short

[11]The dynamic character of this phenomenon is illustrated, for instance, by the ever new discussions in the Cafe Aleph blog (http://blog.marciszewski.eu/) whose main purpose is to provide a platform for the exchange of information on various topics related to the IWV.

methodological commentary[12] to illustrate how the main assumptions of the IWV may be used in the sciences (e.g. in cognitive science).

Thesis 1 (Its philosophical prototype was Aristotle's hylomorphism.)

Every being contains within itself some informational content, which, combined with the material substrate, determines its characteristics (structure, functions, interactions with the world, etc.).

Commentary

The best insight into this content comes with information processing codes, including analogue and digital ones (while the question about the nature of the world: whether it is analogue, digital, or both, remains open). While a given being is dynamic and interacts with the world, the relevant code is a control one; it may therefore be treated as an algorithm code.

Thesis 2 (Accepted e.g. in cognitive science.)

The human mind is an information processing system.

Commentary

When trying to understand the structure and operating mode of this system, it is necessary to construct (ever improved) computational models of the mind, referring to the ever improved (but also: ever developed) data processing systems, including rule-based, connectionist and evolutionary ones. Assuming that in a mathematical description, data processing may be presented as operations on numbers, one may say that mental activity consists in computation (without defining what numbers and operations are involved); and, moreover, that each individual mind has a certain computational power.[13]

[12]The comments provided here could, naturally, be longer—they would then represent a broader discussion of some of the topics related to the IWV. Due to limited space, they need to be succinct, however. A more detailed discussion of this concept can be found in (Stacewicz 2016: 35–47).

[13]The concept of computational power is particularly emphasized by Marciszewski (Marciszewski and Stacewicz 2011: 148).

Thesis 3 (Strongly emphasized by Marciszewski.)

With the development of human civilization, the complexity of problems solved by the mind (through information processing) keeps increasing.

Commentary

A good measure of the complexity of problems is the information processing one, referred to as computational or algorithmic complexity. In the field of theoretical computer science, it is argued that some problems are unsolvable (in practice or in principle) using particular algorithms (e.g. digital ones). This is a strong incentive for asking about the cognitive limitations of the human mind—which might, at its lowest operating level, be an algorithmic system.

To conclude, let us stress that the theses discussed above may be treated as answers to certain fundamental questions, characteristic of all varieties of the IWV. Importantly, however, these questions make it reasonable to accept an informational outlook on the world, and not a specific set of views which is the result of a particular attitude man adopts with respect to the world.

5 A Discussion of Marciszewski's Optimistic Realism

Among the many types of informational worldviews, particular attention should be paid to a view embraced by Marciszewski, which he called 'cognitive optimism'.[14] As declared by the author, the view refers to David Hilbert's and Kurt Gödel's optimism (Marciszewski 2015). Both of them consistently, though in slightly different ways, argued that in principle every mathematical problem must be solved at some point in the development of mathematics (Wang 1996).

The worldview represented by Marciszewski refers directly to Thesis 3 presented in the previous section: 'With the development of human

[14]A more extensive definition provided by Marciszewski is worded as follows: 'realistic optimism about understanding and transforming the world' (Marciszewski and Stacewicz 2011: 224).

civilization, the complexity of problems solved by the mind (through information processing) keeps increasing'.[15] This thesis may be called a worldview one, as by referring to the universal concept of problem it is sufficiently general to include all possible challenges faced by the human civilization. In short: every challenge is about solving some problem.

Consequently, taking the above thesis as a starting point, we may formulate the following general worldview question: 'Is the cognitive power of the human mind able to tackle the growing complexity of problems related to the world (in which man lives)?'. Marciszewski is strongly in favour of a positive (optimist) answer, and the main line of his argumentation is as follows (cf. Marciszewski and Stacewicz 2011: 219–226; Marciszewski 2013, 2019).

Both the history of human civilization and the achievements of individual people prove that the human nature is inherently creatively spontaneous. One of its elementary expressions are new abstract concepts which (despite being elementary) pave the way for new theories and the resulting methods of solving problems. A particularly suggestive example is the idea of zero (one must admit there is hardly anything more fundamental) which has allowed man to develop positional notation, and, subsequently, effective algorithms of arithmetic operations on numbers of any magnitude. Without this idea, mathematics would never be able to overcome the barrier of painfully slow calculations which do not yield to effective automation.[16]

According to Marciszewski, creativity of the human mind—one fruit of which is the concept of zero—is inexhaustible. Even if there is no limit to the complexity of problems the mind encounters in the world (it may be of any magnitude), so there is no limit to the effectiveness of theories and algorithms invented by the mind. This thought is expressed well in the following quotation:

[15]This thesis is an expression of the realism of the standpoint discussed further on: it must be accepted realistically, i.e. in accordance with the actual practice of various sciences, that the complexity of problems they examine keeps increasing.

[16]For example, using the Roman notation (without the zero symbol) it was not possible to develop universal algorithms for addition or multiplication (not to mention more complicated operations; cf. Ifrah 2000).

> Even if there are unsolvable problems in any theory, and thus at every stage of learning about the world (which must be realistically admitted), there is always a chance for it to develop or be replaced with a better theory, so that problems which have been unsolvable until now will become solvable. In this new theory, there will be new unsolvable questions again, but there will always be a chance for *progress*, so we have a perspective for a new move forward. (Marciszewski and Stacewicz 2011: s. 225)

To supplement the above thought with a more computational comment, we may say that the unsolvability of such problems consists in the inability to provide (intersubjectively available) algorithms which would make it possible to effectively solve them within the framework of the theory concerned. In short: the lack of algorithms for problem P proves that the theory is inefficient with respect to P.

Interestingly, however, the unsolvability of certain problems runs very deep. It depends not so much on the high-level structure of the theory, but rather on the elementary operations it permits. In the computational jargon, it is said to depend on the permitted model of computation—the model which determines the particular type of algorithms.[17]

The assertion that there are unsolvable problems for each model of computation (and thus also for the corresponding type of algorithms) is one of the scientific foundations of the IWV. We mentioned this in the comment to Thesis 3. This assertion applies in particular to Turing's model of discrete computations, which has been demonstrated to be unable to define any algorithm to solve the halting problem or the related problems.[18]

---

[17]The most thoroughly investigated and most closely related to information processing in practice is the model of discrete computing, precisely described with the formalism of universal Turing machine (describing data processing by digital computers in an idealized way). It is not the only model, however. Apart from this one, various models of continuous (analogue) computations are examined as well. We will refer to them further on in this paper (Burgin and Dodig-Crnkovic 2013).

[18]As a reminder, the halting problem is expressed in the following question: 'Does a particular (but any) Turing machine determined by its programme and the sequence of input symbols halt after a finite number of steps, or must it run indefinitely?' (cf. Harel 1987).

In view of the above, we must ask about the chance of progress mentioned by Marciszewski. Namely: does a way of overcoming its natural limitations exist for every model of computation, in particular for the Turing model? And if so, what is it?

If we refer to the methodology of computer science to find an answer, it will suggest two ways. The first one may be referred to as a conservative one, as it does not require any change of the model within which we are looking for relevant algorithms. It consists in skilfully dividing the original (unsolvable) problem P into such sub-problems for which, firstly, there exist sufficiently effective local algorithms, and secondly, there exists among these sub-problems a relevant variant of the P. This method is supported by the optimist assumption that while for any generally defined problem P there is no single universal algorithm A which solves it, effective 'dedicated' algorithms can be found for its sub-problems (particularly those we are practically interested in).[19] And their effectiveness—this is yet another topic in Marciszewski's thought—depends on the degree of abstraction of the objects on which algorithms operate (Marciszewski 2006). Usually: the more abstract the object, the more effective the algorithm (for example, certain problems concerning real numbers and functions can be solved more effectively if they are supplanted by complex numbers and functions, more abstract than the former).

Another way of coping with algorithmic unsolvability should be called radical, as it consists in abandoning the model of computation used so far and replacing it with a stronger one. In this stronger model, at least some of the unsolvable problems become solvable. A good example is moving from discrete computations, in fact reducible to a certain type of operations on two distinguishable symbols (e.g. 0 and 1), to continuous-analogue computations (Shannon 1941) which make it possible to operate on values from a certain *continuum* (e.g. the range [0, 1]). This new model is stronger because, theoretically, it ensures solvability of the halting problem in Turing machines (a problem that is unsolvable in the first of the above mentioned models; cf. Mycka and Costa 2004).

[19] This assumption does not need to be true, as we will discuss further on.

Thus, summing up: optimism of the IWV version embraced by Marciszewski is expressed in the belief that the human mind, being a dynamic system, capable of producing new concepts, algorithms, and models of computation, is able to "keep up" with the growing complexity of problems in the world. This belief is supported by the history of human civilization (a creative civilization!) and the findings of computer science methodologists described above.

Does the substantiation presented above rule out any pessimist varieties of the IWV? Are there no compelling reasons to accept that the human mind must reconcile with the insolvability of certain types of problems? Or, to speak more cautiously: with uncertainty about the infallibility of the methods it employs to find solutions?

Let me briefly present three arguments in support of such pessimism. Like the discussion above, they will be grounded in the context of the methodology of computer science.

The first argument is a criticism of the strategy, discussed above, of dividing an unsolvable problem P into sub-problems for which there exist effective local algorithms. It should be noted that for each division of P into particular instances, there must be infinitely many instances and corresponding local algorithms. Otherwise there would exist an algorithm for P which had the form of a simple alternative of a finite number of local algorithms. And in view of the mathematically demonstrated unsolvability of P, simply no such algorithm exists.

Therefore, the human mind cannot be certain that it will be able to find the actual variant which corresponds to the issue concerned among the infinite multiplicity of divisions and sub-problems. Moreover, the original problem P, due to the existence of infinitely many local algorithms related to it, in fact remains unsolved. Its complete solution, one that does not apply to a particular finite set of sub-problems, would require the knowledge and ability to search (in a finite time) through an infinite number of local algorithms. The above objection could be abolished on the condition that the human mind is capable of insight into the comprehensive structure of infinite objects (in this case: sets of sub-problems and the related algorithms). This assumption may appear overly optimistic, however.

The other argument challenging the too far-reaching optimism of IWV advocates concerns the concept of replacing weaker models of computation with stronger ones. Even though this concept is theoretically substantiated (after all, mathematical theories of increasingly stronger models of computation do exist[20]) in practice it may turn out that models stronger than the digital (discrete) one are simply physically unrealisable. This thought is in its essence consistent with the famous Church-Turing thesis, which asserts in one of its variants that all effective (effectively executable) computations can be reduced to digital ones (described strictly using the formalism of the universal Turing machine; cf. Harel 1987; Mycka and Olszewski 2015). If this was the case, we would have to pessimistically conclude that all problems which can theoretically be solved in models stronger than the digital one (e.g. analogue-continuous) in fact must remain unsolvable.

The last of the proposed arguments is independent from the above doubts, and emphasizes the incomplete controllability of algorithms applied by man. This undesirable property describes, for example, the ever more readily used schemes of so-called natural computations (Kari and Rozenberg 2008). These include genetic algorithms, 'imitating' natural evolution, whose constitutive elements include purely random operations on data, such as mutation, crossover, or selection (Michalewicz 1992). It is clear that as soon as random operations are permitted, the authors of algorithms can no longer control them.

Other types of natural computations are taken directly from nature as an element of 'ready-made' systems operating in nature (e.g. bacteria or nervous cells). Even though such systems perform computations which are useful to us, their author is not man, but nature shaped by evolution. Moreover, in many cases man cannot even reasonably separate the algorithm from the 'machinery' by which it is executed.

The incomplete controllability of such algorithms is further magnified if they are characterised by a sufficiently great structural complexity

---

[20]These increasingly stronger models of computation—stronger than the discrete/digital model—are often referred to as models of hypercomputations (cf. Copeland 2002; Ord 2006).

(as a side remark, let us add that even in the case of systems as elementary as cells it is gigantic). Due to the complexity of the system and its controlling algorithm, despite its being used as an element supporting the explanation and forecasting of various phenomena in the world, it calls for an explanatory theory itself. In short: the algorithm works and is effective, but we do not fully understand why. It is a strong argument in favour of pessimism about the possibility of achieving effective control of algorithms and algorithmically-controlled devices.

To conclude the above critical comments, it should be stated that the question about the reasonability of Marciszewski's cognitive optimism remains open. As we have tried to demonstrate above, arguments have their counterarguments, and the latter are not irrefutable. Which shows that when viewing the world from the informational perspective, one may hold various types of the informational worldview.

References

Ajdukiewicz, K. (1934). Naukowa perspektywa świata. *Przegląd Filozoficzny, XXXVII*, 409–416.

Ajdukiewicz, K. (2003). *Zagadnienia i kierunki filozofii*. Kęty and Warszawa: Wydawnictwo Antyk—Fundacja Aletheia.

Ajdukiewicz, K. (2006). *Język i poznanie* (Vol. 1). Warszawa: PWN.

Burgin, M., & Dodig-Crnkovic, G. (2013). *Typologies of computation and computational models*. arXiv:1312.2447[cs].

Cafe Aleph (http://blog.marciszewski.eu/), an academic discussion blog of W. Marciszewski and P. Stacewicz.

Copeland, J. (2002). Hypercomputation. *Mind and Machines, 12*, 461–502.

Denning, P. J. (2005). Is computer science science? *Communications of the ACM, 48*(4), 27–31.

Harel, D. (1987). *Algorithmics: The spirit of computing*. Boston, MA: Addison-Wesley.

Ifrah, G. (2000). *The universal history of numbers: From prehistory to the invention of the computer*. New York: Wiley.

Ito, Y. (1991). Birth of Joho Shakai and Johoka concepts and their diffusion outside Japan. *Keio Communication Review, 13*, 3–12.

Kari, L., & Rozenberg, G. (2008). The many facets of natural computing. *Communications of the ACM, 10*(51), 72–83.

Knuth, D. E. (1974). Computer science and its relation to mathematics. *American Mathematical Monthly, 4*(81), 323–343.

Marciszewski, W. (1996). *Sztuka Dyskutowania*. Warszawa: Wydawnictwo „Aleph".

Marciszewski, W. (2006). The Gödelian speed-up and other strategies to address decidability and tractability. *Studies in Logic, Grammar and Rhetoric, 22*, 9–29.

Marciszewski, W. (2013). Racjonalistyczny optymizm poznawczy w Gödlowskiej wizji dynamiki wiedzy. In R. Ziemińska (Ed.), *Przewodnik po epistemologii*. WAM: Kraków.

Marciszewski, W. (2015). On accelerations in science driven by daring ideas: Good messages from fallibilistic rationalism. *Studies in Logic, Grammar and Rhetoric, 53*, 19–41.

Marciszewski, W. (2019). The progress of science from a computational point of view: The drive towards ever higher solvability. *Foundations of Computing and Decision Sciences, 44*, 11–26.

Marciszewski, W., & Stacewicz, P. (2011). *Umysł – Komputer – Świat. O zagadce umysłu z informatycznego punktu widzenia*. Warszawa: Akademicka Oficyna Wydawnicza EXIT.

Michalewicz, Z. (1992). *Genetic algorithms + Data structures = Evolution programs*. Berlin: Springer.

Murawski, R. (2014). *Filozofia informatyki. Antologia*. Poznań: Wydawnictwo Naukowe UAM.

Mycka, J., & Costa, J. F. (2004). Real recursive functions and their hierarchy. *Journal of Complexity, 20*, 835–857.

Mycka, J., & Olszewski, A. (2015). Czy teza Churcha ma jeszcze jakieś znaczenie dla informatyki? In P. Stacewicz (Ed.), *Informatyka a filozofia. Od informatyki i jej zastosowań do światopoglądu informatycznego* (pp. 53–74), Warszawa: Oficyna Wydawnicza PW.

Ord, T. (2006). The many forms of hypercomputation. *Applied Mathematics and Computation, 178*, 8–24.

Shannon, C. (1941). Mathematical theory of the differential analyzer. *Journal of Mathematics and Physics MIT, 20*, 337–354.

Stacewicz, P. (2016). Informational worldview: Scientific foundations and philosophical perspectives. *Studies in Logic, Grammar and Rhetoric, 47*(60), 35–47.

Tatarkiewicz, W. (1990). *Historia filozofii* (Vols. 1–3). Warszawa: PWN.
Turing, A. M. (1936). On computable numbers, with an application to the Entscheidungsproblem. *Proceedings of the London Mathematical Society, 42,* 230–265.
Turing, A. M. (1950). Computing machinery and intelligence. *Mind, 49,* 433–460.
Twardowski, K. (1965). *Wybrane pisma filozoficzne.* Warszawa: PWN.
Wang, H. (1996). *Logical journey: From Gödel to philosophy.* Cambridge and London: MIT Press.
Żurada, J. M. (1992). *Introduction to artificial neural systems.* St. Paul: West Publishing Company.

Index

A

abstraction
- abstraction operator 195
- acts of abstraction 187
- axiom of abstraction (AA) 188, 190, 191, 194, 195
- level of abstraction 187, 188, 254
- order of abstraction 188
- processes of abstraction 180

abstracts
- ontological types of abstracts 190
- variables ranging over abstracts 185

act
- act of perception 126–128, 130, 132
- category of act 5, 72
- complex act 126
- independent act 126, 127
- moral judgement of act 72
- notion of act 72, 73
- psychological act 121, 124–126, 138
- thetical act 126

Adler, Alfred 8, 143, 160–163, 168, 172
Adorno, Theodor 172
Ajdukiewicz, Kazimierz 2, 3, 8, 10–12, 22–24, 27–30, 32, 42, 96–98, 101, 103, 104, 106, 133, 264, 266
algebraic topology 198
Annales School 172
apriorism 66
Aristotle 65, 66, 78, 100, 123, 189, 190, 247–249, 271
arithmetics 197, 200
Auerbach, Walter 25, 29, 148

A. Drabarek et al. (eds.), *Interdisciplinary Investigations into the Lvov-Warsaw School*, History of Analytic Philosophy,
https://doi.org/10.1007/978-3-030-24486-6

Augustine, Saint 249, 250
axiology
  axiology project 4
  axiomatization 199
  Czeżowski's axiology 69
  empirical axiology 64
  realistic axiology 64
axiomatization
  axiomatization of finitely many-valued propositional logics 9, 10
  axiomatization of philosophy 111
axiom system 206–208, 211, 213
  axiom systems for finitely many-valued logics 205

B

Bacon, Francis 179
Bad, Hersch 23, 29
Baley, Stefan 7, 23, 29, 95, 128, 137, 141, 144, 146, 147
Bandrowski, Bronisław 23
Bar-Hillel, Yehoshua 192
Baron, Alfons 39
Bartholomew of Jasło 47
Batóg, Tadeusz 42, 44
Będkowski, Marcin 10, 11
Bell, Daniel 265
Benedykt Hesse of Krakow 47
Berezowska, Maja 40, 41
Błachowski, Stefan 7, 23, 141, 143, 146, 147
Blaustein, Leopold 25, 29, 38, 148
Bocheński, Józef M. 25–28, 30, 192
Bogacki, Feliks 50
Bornstein, Benedykt 26
Borowski, Marian 23, 72
Brentano, Franz 6, 18, 27, 28, 121, 123, 125–127, 132, 141, 149, 152, 154, 164, 170
Brożek, Anna 5, 233
Bunge, Mario 252, 253

C

Carnap, Rudolf 22, 233
categoremata 192
categorial grammar (CG) 188, 191, 192
Chrzanowski, Ignacy 38
Chwistek, Leon 26
Citlak, Amadeusz 7, 8, 167, 169, 171, 173
computer science
  apparatus of computer science 12
  concepts of computer science 267
  language of computer science 267
  methodology of computer science 275, 276
  theoretical computer science 267, 272
Comte, Auguste 243
connectives
  propositional connectives 193, 206–213
  standard connectives 10, 209, 211, 212
  unary connectives 209
conventionalism 104
Cracow Circle 25
cratism
  cratism theory 160, 161, 170, 171
Cresswell, Max 192

Csató, Edward 25, 29
Czerny, Zygmunt 24
Czeżowski, Tadeusz 2–4, 23, 25, 27–30, 42, 51–61, 66–71, 79, 81, 82, 98, 101–103, 130, 137

D

Dąmbska, Izydora 22, 25, 26, 28–30, 95, 97, 98, 104, 105
Davidson, Donald 136, 137
deontology 3, 49
Descartes, René 36, 120–123, 250
Dickstein-Wieleżyńska, Julia 38
Dorpat School of the Psychology of Religion 8, 160, 163, 168
Drabarek, Anna 4, 5
Drewnowski, Jan 24, 25, 28

E

Elzenberg, Henryk 26, 42
Epstein-Weinberg, Sabina 38
ethics
  ethics for academic professors 48
  ethics in education 48
  ethics of scholarly work and academic teaching 49
  ethics of scholars 47
eudaemonism 70
extrospection 121

F

felicitology 72
Festinger, Leon 8, 143, 160, 166, 167
Filozofówna, Irena 29
Fleck, Ludwik 26, 43
Fredkin, Ed 265
Frege, Gottlob 27, 184, 189, 192
Frenkel, Karol 4, 5, 72, 74
Frydman, Sawa 25

G

Ganszyniec, Ryszard 24
Geach, Peter 192
Giedymin, Jerzy 234
Ginsberg-Blaustein, Eugenia 25, 38
Gödel, Kurt 9, 187, 196–200, 226, 246
Goodman, Nelson 181, 186
Grabowski, Tadeusz 38
Gralewski, Jan 38
Grzegorczyk, Andrzej 9, 29, 180, 186, 188, 190, 202

H

Handelsman, Marceli 39
happiness 5, 64, 65, 67, 70, 72, 77, 82, 83
Hejmej, Andrzej 240, 242
Heller, Michał 245, 246
Helm, Sarah 41, 42
Herling-Grudziński, Gustaw 37
Hilbert, David 199, 272
Hiż, Henryk 24, 29, 30, 192
Höfler, Alois 124, 147
Holocaust 3, 35–37, 39, 42, 43
Hosiasson-Lindenbaum, Janina 24, 28, 29, 32, 38
Hume, David 4, 36

Husserl, Edmund 18, 27, 126, 127, 132, 192, 247

I

Igel, Salomon 23, 29
implication 82, 193, 240
Ingarden, Roman 17, 18, 26, 27, 30, 97, 126, 127, 132
Ingel, Salomon 39
interdisciplinarity 12, 13, 239–242, 251–255
interdisciplines 252, 253
intersubjectivity 135
intertextual relationships 88, 93, 94, 111
introspection 5, 7, 73, 121, 130, 142, 144, 149–152, 155, 166, 168, 169
Irzykowski, Karol 39
Ivanyk, Stepan 6, 29

J

Jadacki, Jacek J. 22, 232, 233, 235, 245
Janiszewski, Zygmunt 24
Jan of Ludzisko 47
Janowski, Jarosław Maciej 11, 12
Jaśkowski, Stanisław 24
Jaxa-Bykowski, Mieczysław 23
Jesus Christ 8, 160, 161, 164, 167, 169, 170, 172
John II Casimir 19
Jordan, Zbigniew 25, 28, 30
Józef Mianowski Fund 55
judgment/s
  affirmative 124, 127
  analytical 65, 122
  class of 121, 122
  cognitive 77
  empirical 70, 164, 266
  ethical 66, 71, 73, 74, 165
  existential 124, 126
  individual 69, 70
  logical 6, 138, 164, 226, 227
  moral 72–74, 77, 79
  negative 121, 131
  perceptual 122, 126, 131–137
  positive 77, 121
  psychological 164

K

Kaczorowski, Stanisław 23
Kamińska, Janina 39, 41–44
Kant, Immanuel 111, 247, 248
Karski, Jan 40, 43
Kiedrzyńska, Wanda 39–41
Kierski, Aleksander 39
Kleiner, Juliusz 24
Klein, Julie Thompson 240
Koch, Julius 120
Kodi, Józefa 38
Kokoszyńska-Lutmanowa (Kokoszyńska), Maria 25, 28, 30
Kołaczkowski, Stefan 38
Kołakowski, Leszek 243
Kołłątaj, Hugo 49, 50
Konarski, Stanisław 49
Korcik, Antoni 25, 30
Korporowicz, Leszek 240
Kostelnyk, Gabriel 29
Kotarbińska, Janina 24, 39, 41–44, 181

Kotarbiński, Tadeusz 2, 4, 5, 9–11, 21, 23, 24, 28–30, 32, 39, 66, 71, 72, 75–79, 81, 88, 96–98, 101, 102, 106, 107, 109, 131, 180, 181, 183–188, 190, 192, 217, 218, 220–224, 234, 235, 248
kratism 8, 23, 99, 143, 145
  kratism theory 8, 23, 99, 143, 145
Kreutz, Mieczysław 7, 23, 142, 144, 147, 148
Kridl, Manfred 24
Külpe, Oswald 132, 150
Küng, Guido 185
Kuryłowicz, Jerzy 24

L

Lambek, Joachim 192
Lam, Stanisław 94
Lanckorońska, Karolina 40
Lejewski, Czesław 24, 30
Łempicki, Jan 39
Łempicki, Stanisław 24, 52, 94
Łempicki, Zygmunt 24, 29, 39
Leśniewski, Stanisław 2, 3, 9, 23, 24, 29, 32, 88, 98, 100, 101, 106, 107, 109, 180, 181, 185–187, 192
Lindenbaum, Adolf 24, 29, 38
Lloyd, Seth 265
logic
  ancient 189
  elementary 97
  formal 27, 28, 101
  language of 102, 189
  many-valued 3, 10, 32, 108, 206, 208–211, 213
  mathematical 2, 24, 25, 27, 99, 110
  modern 102
  propositional 10, 205, 206, 208–210, 213
  quantification 185
  standard 185, 206, 208–211
  symbolic 184, 189
  three-valued 9, 10, 205, 207, 208, 213
logical operators 192
logical types 191
Łukasiewicz, Jan 2, 3, 10, 23, 24, 27, 29, 30, 32, 88, 95, 96, 98–100, 106–111, 166, 192, 205, 206, 211–213
Łuszczewska-Romahn, Seweryna 25, 39, 41–44
Lvov-Warsaw School (LWS)
  ethicists of the 72, 75, 78
  philosophers of the 2, 3, 37, 38, 41, 63, 65, 66, 77–80
  representatives of the 3, 4, 29, 74, 88, 98, 241, 242, 245, 261
  women in the 39

M

Mączka, Janusz 245, 246
Mahrburg, Adam 50
Makowski, Szymon Stanisła 49
Marciszewski, Witold 8, 9, 12, 13, 180, 181, 259, 260, 262, 271–273, 275, 276, 278
Markin, Estera 38, 143, 161, 163
Massey, William S. 198
Matellmann, Joachim 38

mathematics 1, 2, 24, 27, 49, 100, 183, 185, 198–200, 241, 252, 266, 272, 273
McDowell, John 136
Meinong, Alexius 27, 164, 165, 184
Milbrandt, Mieczysław 24, 29, 39
Mizińska, Jadwiga 239
Modrzewski, Andrzej Frycz 49
monodisciplines 252, 253
Montague, Richard 192
Moore, George E. 27
Morsdorf, Jan 39
Mostowski, Andrzej 24
multidisciplinarity 12, 240, 251, 252, 255

N

Natorp, Paul 120, 123
negation 93, 109, 195, 226
Niedźwiedzka-Ossowska, Maria 24
nominalism
  extreme 189, 190, 193, 194, 197, 199
  Kotarbiński's 185, 187, 188, 190
  moderate 9, 188–190, 201
  pragmatic 9, 180, 189, 202
  radical 9
  reistic 9, 183, 186
Nowiński, Czesław 25

O

Ochorowicz, Julian 50
Olech, Adam 126–128, 132
Oleksyuk, Stepan (Oleksiuk Stefan) 29
Ossowski, Stanisław 24, 25, 28–30, 172

P

Paczkowska, Elżbieta 120
Pakszys, Elżbieta 3, 39, 42
Panenkowa, Irena 39
Pański, Antoni 24, 29, 38
Paprzycka, Katarzyna 233
paraphrase/s
  reistic 11, 218, 221, 222, 224, 225, 235
  semantic 11, 234
  theory of 11, 225
paraphrastic
  premises 11, 218, 234
  relation 228, 232, 234
  relationship 10, 218
Paul of Worczyn 47
Peano, Giuseppe 182
Peirce, Charles Sanders 200
Pelc, Jerzy 22, 222
perception
  act of 126–128, 130, 132
  external 121, 129–132, 149
  false 129
  internal 121, 122, 129, 130, 149
  psychology of 148
  sensory 180, 184, 200
  structure of 6, 11, 129–131, 137
  syncretic 144
perceptron 267
Perzanowski, Jerzy 97
phenomenology 18, 245
philosophy
  analytic 1, 10, 27, 32, 217, 234
  Christian 38, 262
  general 3, 12, 28, 245
  philosophy in Poland 31, 32, 243
  philosophy of idealism 234
  philosophy of language 2, 26, 27, 244

philosophy of nature 22, 241, 248, 261
philosophy of science 2, 11–13, 19, 23, 27, 43, 142, 180, 241, 245, 248, 249, 252, 261, 262, 267
Polish 1, 2, 18, 20, 21, 23, 27, 31, 145, 243, 262
scientific 22, 32, 110, 241, 242, 244, 245, 248–250
Piłsudski, Józef 41
Plato 8, 9, 23, 36, 78, 99, 110, 169–171, 180, 189
Platonism 180, 181, 189, 190
pragmatic 8, 190
Poczobut, Robert 50, 240, 241, 251–255
Poincaré, Henri 27, 247
Popławski, Stanisław 49
Posmysz, Zofia 43
Post, Emil 10, 210
Poznański, Edward 24, 28, 30
praxeology 72, 75, 77, 102
Presburger, Mojżesz 24, 29
Przełęcki, Marian 22, 217–222, 224
psychology 1, 2, 6–8, 19, 21, 23, 102, 104, 106, 108, 110, 128, 130, 138, 141–149, 152–154, 159, 160, 163, 164, 167–170, 172, 173
Putnam, Hilary 199

Q
quantifiers 183, 185, 186, 188, 192, 193
Quine, Willard Van Orman 8, 181, 183, 185, 190, 199, 200

R
Radzki, Mateusz M. 9, 10, 206–208, 211
Rajgrodzki, Jakub 24, 29, 38
Rasiowa, Helena 29
rationalism
epistemological 200
extreme 189
pragmatic 8, 9, 179–181, 190, 198, 200, 202
realism
extreme 248
metaphysical 127
moderate 248
optimistic 272
pragmatic 8
reism
Kotarbiński's reism 10, 188
liberal reism/resim in a liberal sense 9, 202
metaphysical 101
semantic 11, 96, 101, 218–221
thesis of 10, 11, 217, 219–221
reistic sentence 11, 219–221, 223
Rej, Mikołaj 32
relativism 66, 67, 102, 104
non-rational relativism 102
relativism in ethics 66
representation/s 19, 101, 120, 121, 123, 125, 127, 128, 130–132, 145, 148, 165, 200
abstract 122
class of 121
combination of 123
Rickert, Heinrich 234
Ritter, Alexander 36
Romahn, Edmund 39
Rosch, Friedrich 36

Rosser, John B. 10, 206, 208, 209
Russell, Bertrand 27, 184, 191, 199
Rutski, Jan 25, 29
Ryś, Zofia 40
Rzepa, Teresa 6, 7, 142, 144, 145, 147, 148, 151, 153, 161, 166, 167, 170

S

Salamucha, Jan 25, 28, 29, 39
Scheler, Max 67
Schetz, Adriana 136, 137
Schmierer, Zygmunt 25, 27–29
semiotics 19, 23, 27, 28
Sendlerowa, Irena 43
Simons, Peter 185, 186
Siwicki, Jerzy 37
Słupecki, Jerzy 10, 24, 205, 207, 208, 213
Smolka, Franciszek 23
Śniadecki, Jan 49, 50
Sobociński, Bolesław 24, 25, 29, 30
Socrates 8, 36, 41, 66, 97, 143, 160, 170–172, 189
Sośnicki, Kazimierz 23, 78
Spasowski, Władysław 39
Stacewicz, Paweł 12, 13, 260, 268, 269, 271, 272
Stanisław of Skarbimierz 47, 48
Staszic, Stanisław 49, 50
Stawarski, Adam 39
Strauss, Richard 35
Suszko, Roman 181, 197
Świętochowski, Aleksander 50
Swieżawski, Stefen 25, 28
syncategoremata 192, 193
Szaniawski, Klemens 22, 29, 97
Sztejnbarg, Dina 24, 39, 43
Szymon Marycjusz of Pilzno 48

T

Taine, Hippolyte 124
Tarski, Alfred 2, 3, 8, 9, 24, 27, 30, 32, 88, 93, 181, 186, 192, 233
Tatarkiewicz, Władysław 4, 5, 12, 17–21, 23, 24, 28–30, 65–68, 71–73, 77, 79, 81, 82, 261
Tec, Nechama 43
Tenner-Gromska, Daniela 23, 37, 39
thetical act 126
thetical function 124
thetical moment 126, 127
transdisciplinarity 11, 12, 240, 254, 256
transdisciplines 252
Treter, Mieczysław (Michał) 23, 29, 39
Trzebiecki, Andrzej 49
Turek, Józef 242
Turing model 13, 268, 275
Turquette, Atwell R. 10, 206, 208, 209
Twardowski, Kazimierz 1–7, 12, 17–26, 28–31, 37, 39, 41, 51–54, 57–61, 64–68, 71–74, 78, 79, 81, 88, 94–111, 119–128, 130, 132, 136–138, 142–150, 152–156, 159, 164, 218, 243–245, 259, 261, 263
Tyburski, Włodzimierz 3, 4

U

Umesao, Tadao 265

V

van Benthem, Johan 192
Vienna Circle 22, 27, 32, 243

W

Wajsberg, Mordchaj 10, 24, 29, 206, 213
Wallis-Walfisz, Mieczysław 24
Wang, Hao 196, 272
Warsaw School of Logic 2, 24, 25, 27, 31, 99, 111
Wasilewski, Michał 38
Weigl, Rudolf 43
Weryha, Władysław 21
William of Ockham 182
Wiśniewski, Antoni 49
Witkiewicz, Stanisław Ignacy (Witkacy) 39
Witwicki, Władysław 2–5, 7, 8, 19–21, 23–25, 71–74, 79, 81, 82, 97–99, 106, 109–111, 128–130, 137, 141, 143–145, 147, 149, 151–154, 159–173
Wójtowicz, Anna 223, 235
Woleński, Jan 2, 26, 28, 93, 95, 144, 206, 213, 226, 241, 243–245
World War I 2, 8, 52
World War II 1–3, 22, 25, 30, 37, 40, 43, 76, 148
Wundheiler, Aleksander 24, 28, 30
Wundt, Wilhelm 6, 8, 141, 145, 147, 150, 152, 160, 169
Würzburg School 150

Z

Zajkowski, Józef 25, 29
Zarycki, Miron 29
Zawirski, Zygmunt 2, 11, 12, 23, 25, 27–30, 241, 244–251, 255, 256
Zermelo, Ernst 187, 188
Zygmuntowicz, Itka Frajman 43

Printed by Printforce, the Netherlands